The Civic Potential of Video Games

This report was made possible by grants from the John D. and Catherine T. MacArthur Foundation in connection with its grant making initiative on Digital Media and Learning. For more information on the initiative visit www.macfound.org.

The Civic Potential of Video Games

Joseph Kahne, Ellen Middaugh, and Chris Evans

The MIT Press
Cambridge, Massachusetts
London, England

For information about special quantity discounts, please email special_sales@mitpress.mit.edu.

This book was set in Stone Serif and Stone Sans by the MIT Press. Printed and bound in the United States of America.

Library of Congress Cataloging-in-Publication Data

Kahne, Joseph.
The civic potential of video games / Joseph Kahne, Ellen Middaugh, and Chris Evans.
    p. cm.—(The John D. and Catherine T. MacArthur Foundation reports on digital media and learning)
Includes bibliographical references.
ISBN 978-0-262-51360-9 (pbk. : alk. paper)
1. Video games—Social aspects—United States. 2. Video games and teenagers—United States. 3. Youth—United States—Political activity. 4. Youth—Social networks—United States. I. Middaugh, Ellen. II. Evans, Chris, 1981– III. Title.
GV1469.34.S52K34   2009   794.8—dc22   2009007499

10 9 8 7 6 5 4 3 2 1

Contents

Series Foreword

The John D. and Catherine T. MacArthur Foundation Reports on Digital Media and Learning, published by the MIT Press, present findings from current research on how young people learn, play, socialize, and participate in civic life. The Reports result from research projects funded by the MacArthur Foundation as part of its $50 million initiative in digital media and learning. They are published openly online (as well as in print) in order to support broad dissemination and to stimulate further research in the field.

Acknowledgments

The authors would like to thank Craig Wacker, Connie Yowell, and Benjamin Stokes at the MacArthur Foundation; the scholars and researchers who gave us feedback on the survey instrument, the report, and the research arena as a whole: Craig Anderson, Sasha Barab, Linda Burch, Lance Bennett, Brad Bushman, David Chen, Seran Chen, Rana Cho, Connie Flanagan, Jim Gee, Eszter Hargittai, Betty Hayes, Mimi Ito, Henry Jenkins, Barry Josephs, Scott Keeter, Miguel Lopez, Ryan Patton and Smithsonian Summer Camps, Rebecca Randall, Chad Raphael, Katie Salen, Rafi Santos and Global Kids, David W. Shaffer, Constance Steinkuehler, Doug Thomas, and Dmitri Williams.

We are especially grateful to Amanda Lenhart, Lee Rainie, Alexandra Rankin Macgill, and Jessica Vitak of the Pew Internet and American Life Project and to Sydney Jones, Pew Internet research intern for collaborating on the Pew Teens, Video Games, and Civics Survey. The data analysis and findings presented in that report are central to much of the analysis presented here. The authors are solely responsible for all conclusions.

About This Report

This report draws from the 2008 Pew Teens, Video Games, and Civics Survey, a national survey of youth and their experiences with video games done in partnership with Amanda Lenhart at the Pew Internet and American Life Project, with funding from the John D. and Catherine T. MacArthur Foundation. That survey led to the report, "Teens, Video Games, and Civics," which examines the nature of young people's video game play as well as the context and mechanics of their play. In addition to examining the relationship between gaming and youth civic engagement, "Teens, Video Games, and Civics" also provides a benchmark for video and online gaming among young people on a national level and the first broad, impartial look at the size and scope of young people's general gaming habits.

This current report, *The Civic Potential of Video Games*, focuses solely on the civic dimensions of video game play among youth. Although it shares some text and findings with the "Teens, Video Games, and Civics" report, it provides a more detailed discussion of the relevant research on civics and gaming. In addition, this report discusses the policy and research implications of these findings for those interested in better understanding and promoting civic engagement through video games. The interpretation of data and the discussion of implications reflect only the authors' perspectives. The Pew Internet Project and the MacArthur Foundation are nonpartisan and take no position for or against any technology-related policy proposals, technologies, organizations, or individuals and do not take a position on any of the proposals suggested here.

About the Civic Engagement Research Group (CERG)

CERG is a research organization based at Mills College in Oakland, California, that conducts quantitative and qualitative research on youth civic engagement. The group looks at the impact of civic learning opportunities and digital media participation on young people's civic capacities and commitments, as well as civic opportunities and outcomes in public schools. The goal is to develop an evidence base on effective civic education practices and policies. Joseph Kahne is currently the Abbie Valley Professor of Education, Dean of the School of Education at Mills College, and CERG's Director of Research. Ellen Middaugh is Senior Research Associate at CERG. Chris Evans is Senior Program Associate at CERG.
www.civicsurvey.org.

About Princeton Survey Research Associates (PSRA)

PSRA conducted the survey that is covered in this report. PSRA is an independent research company specializing in social and policy work. The firm designs, conducts, and analyzes surveys worldwide. Its expertise also includes qualitative research and content analysis. With offices in Princeton, NJ, and Washington, DC, PSRA serves the needs of clients around the nation and the world. The firm can be reached at 911 Commons Way, Princeton, NJ 08540, by telephone at 609-924-9204, or by email at ResearchNJ@PSRA.com.

The Civic Potential of Video Games

The Civic Dimensions of Video Games

In Pew's Teens, Video Games, and Civics Survey, we asked 1,102 youth ages 12 to 17 if they had played a video game. Only 39 said no.[1] We found that nearly one-third of all 12- to 17-year-olds report playing video games every day or multiple times each day, and three-fourths report playing at least once a week.

The games youth play are diverse. Indeed, in our survey, we classified 14 different genres of games that youth play. Eighty percent of youth play games from more than five different genres. These genres range from sports games (for example, the *Madden* series), to playing music (*Guitar Hero*), to first-person shooter games (*Halo*), to more civically oriented games (*Civilization*). Some games have violent content, but by no means all. Almost all youth who play games that contain violent content also play games that do not.[2]

Youth play these games on computers, game consoles, portable gaming devices, and cell phones. They play alone, with others online, with friends in the room, as part of a team or guild, in school, supervised, and unsupervised. In addition, many game-related activities arise around game play (what Ito et al. refer to as "augmented play"[3]), including visiting and contributing to Web sites about specific games, participating in chat rooms about the game, and customizing the gaming experience by developing and using "cheats" and "mods."[4]

In short, video games are now a very significant part of young people's lives. But in what ways? Although we know that young people play games frequently, the relationship of this activity to adolescent development has not been fully explored.

Over the years, as game design has become more sophisticated and the content more varied, debates over the value of games have surfaced. Media watchdog groups such as the National Institute on Media and the Family warn that video games can lead to social isolation, aggressive behavior, and reinforced gender stereotypes.[5] Advocates of video games' potential, on the other hand, call attention to the "tremendous educative power" of games to integrate thinking, social interaction, and technology into the learning experience.[6] Digital media scholars such as Henry Jenkins also highlight how video games and other forms of digital media can foster "participatory cultures" with "relatively low barriers to artistic expression and civic engagement."[7]

Although public debates often frame video games as either good or bad, research is making it clear that when it comes to the effects of video games, it often depends. Context and content matter.

To date, the main areas of research have considered how video games relate to children's aggression and to academic learning.[8] However, digital media scholars suggest that other social outcomes also deserve attention. For example, as games become more social, some suggest they can be important spheres in which to foster civic development.[9] Others suggest that games, along with other forms of Internet involvement, may take time away from civic and political engagement.[10] No large-scale national survey, however, has yet examined the civic dimensions of video games. Given the ubiquity of video game play among youth, this is a serious omission. Levels of teen civic engagement are lower than desired, adolescence is a time when the development of civic identity is in full force, and, as noted above, video game play has been described both as a means of fostering civic engagement and as a force that may undermine civic goals. In an effort to bring data to bear on this debate, we draw on data from the Pew Teens, Video Games, and Civics Survey. This nationally representative survey of youth ages 12 to 17 enables us to examine the relation between young people's video game play and their civic and political development.

Youth Civic and Political Engagement

In his book *Democracy and Education*, noted philosopher and educational reformer John Dewey argued that we must not take for granted the formation of the habits and virtues required for democracy. He believed these must be developed by participating in democratic communities—those places where groups of individuals join together around common interests and where there is "free and full interplay" among those holding differing views. Democratic communities were also characterized by dialogue and active experimentation that reflected social concerns.[11]

Many others have since adopted Dewey's perspective that this kind of robust community participation is fundamental to the health of a democratic society. To have a government and society that fairly represent and support diverse and sometimes competing needs requires a nation of what Benjamin Barber calls, "small *d* democrats"—citizens who participate at multiple levels both individually and collectively.[12] This includes formal political activities such as voting and informal civic activities such as volunteering, working with others on community issues, and contributing to charity. Sustained, lifelong participation requires a strong sense of commitment to civic engagement, an informed interest in the political and civic issues that affect one's community and country, and a willingness to take action to address local and national problems.

Unfortunately, levels of civic engagement are lower than desirable, most evidently among the young. The Center for Research on Civic Learning and Engagement found that 58 percent of youth aged 15 to 25 were "disengaged," defined as participating in fewer than two types of either electoral (voting, wearing a campaign button, signing an email or written petition) or civic (volunteering, raising money for charity) activities.[13] On the 2006 National Assessment of Educational Progress (NAEP) Civics Assessment, only 9 percent of high school seniors could list two ways a democratic society benefits from citizen participation.[14]

Such disengagement is not confined to youth. A panel of experts convened by the American Political Science Association recently found that "citizens participate in public affairs less frequently, with less knowledge, and enthusiasm, in fewer venues, and less equitably than is healthy for a vibrant democratic polity."[15] Clearly, democratic engagement is not guaranteed. Rather, it must be nurtured in each successive generation of young people.

Developmental psychologists suggest that adolescence is an important time for such nurturing to begin because it is a time when youth are thinking about and trying to anticipate their lives as adults and when they are working to understand who they are and how they will relate to society.[16] As Erik Erikson noted, it is a critical time for the development of sociopolitical orientations.[17] Therefore, it is important to assess the extent to which young people are experimenting with the civic and political activities available to them and developing commitments to future participation.

Potential Links between Video Games and Youth Civic and Political Development

Gaming may foster civic engagement among youth. Several aspects of video game play parallel the kinds of civic learning opportunities found to promote civic engagement in other settings. Simulations of civic and political action, consideration of controversial issues, and participation in groups where members share interests are effective ways, research finds, for schools to encourage civic participation.[18] These elements are common in many video games. In addition, many games have content that is explicitly civic and political in nature. *SimCity*, for example, casts youth in the role of mayor and requires that players develop and manage a city. They must set taxes, attend to commute times, invest in infrastructure, develop strategies for boosting employment, and consider their approval rating (see box 1 on page 16 for an example of *SimCity* in action).

Furthermore, interactions in video games can model Dewey's conception of democratic community—places where diverse groups of individuals with shared interests join together, where groups must negotiate norms, where novices are mentored by more experienced community members, where teamwork enables all to benefit from the different skills of group members, and where collective problem solving leads to collective intelligence.

Henry Jenkins, a leading scholar in the digital media field, has highlighted the potential of the participatory cultures that arise through engagement with digital media.[19] These participatory cultures support communities of shared interests within which participants create and share what they create with others. Those with more experience also mentor others. According to Jenkins, the new participatory culture created by video games and other forms of digital media

offers many opportunities for kids to engage in civic debates, to participate in community life, to become political leaders—even if sometimes only through the "second lives" offered by massively multiplayer games or online fan communities. Here, too, expanding opportunities for participation may change their self perceptions and strengthen their ties with other citizens. Empowerment comes from making meaningful decisions within a real civic context: we learn the skills of citizenship by becoming political actors and gradually coming to understand the choices we make in political terms. . . . The step from watching television news and acting politically seems greater than the transition from being a political actor in a game world to acting politically in the real world.[20]

Doug Thomas and John Seely Brown make a similar point in their discussion of virtual worlds. "The dispositions being developed in *World of Warcraft*," they write,

are not being created in the virtual and then being moved to the physical, they are being created in both equally. . . .

Players are learning to create new dispositions within networked worlds and environments which are well suited to effective communication, problem solving, and social interaction.[21]

For example, players of *World of Warcraft* generally join or form guilds. As members of these associations, they plan and carry out coordinated raids against the enemy. They recruit new members and train them, as well as resolve conflicts between guild members and establish an explicit or implicit code of conduct.[22]

Dewey, writing at the beginning of the twentieth century, wanted schools and classrooms to prepare youth for democracy by creating "miniature communities" that simulated civic and democratic dynamics. Youth would experience democratic life at the same time that they developed related skills.[23] At the beginning of the twenty-first century, those designing and studying video games are making similar claims about their potential. It therefore makes sense to ask whether video games support or constrain the pursuit of democratic goals.

Research Questions

The Pew Teens, Video Games, and Civics Survey, the first large study with a nationally representative sample of youth, sheds light on relationships between video game play and civic engagement by measuring the quantity, civic characteristics, and social context of gaming. It explores, in addition, the relationship between the civic characteristics and social context of game play, on one hand, and varied civic outcomes, on the other. In this report, we use the results of this survey to examine how teens' exposure to these civic gaming experiences relates to their civic participation. We define video games as any type of interactive entertainment software, including any type of computer, console, online, or mobile game.

Specifically, we consider:

The Quantity of Game Play   Do teens who play games every day or for many hours at a time demonstrate less or more commitment and engagement in civic and political activity? Do they spend less or more time volunteering, following politics, protesting?

The Civic Characteristics of Game Play   Do teens who have civic experiences while gaming—such as playing games that simulate civic activities, helping or guiding other players, organizing or managing guilds (an opportunity to develop social networks), learning about social issues, and grappling with ethical issues—demonstrate greater commitment to and engagement in civic and political activity than those with limited exposure to civic gaming experiences?

The Social Context of Game Play   Do teens who play games with others in person have higher levels of civic and political engagement than those who play alone? Does playing games with others online have the same relationship to civic engagement as playing games with others in person? How often do youth have social interactions around the games they play, for example participating in online discussions about a game? How do these interactions relate to civic and political engagement?

The Demographic Distribution of Civic Gaming Experiences   Do factors such as gender, family income, race, and ethnicity influence the frequency of civic gaming experiences that members of these groups have? Do certain games provide more of these experiences than others?

Why Study the Quantity of Video Game Play?

Much of the public discourse around game play concerns whether the amount of time youth spend playing "video games" is good or bad. These broad statements do not make meaningful distinctions between the characteristics of particular games or the social context in which they are played. We therefore ask whether the overall quantity of video game play is related to civic and political engagement before considering how the characteristics and context of game play might relate to civic engagement.

Our interest in these questions also reflects analyses that suggest that spending significant time playing video games could lessen the time youth have to spend participating in civic and political life. Indeed, Nie and colleagues found that after controlling for education and income, heavy Internet use was associated with less face-to-face contact with friends, families, and neighbors, particularly when participants used the Internet at home rather than solely at work.[24] In a related argument, Robert Putnam notes that what were previously social leisure activities, such as card games, have now been largely replaced by electronic versions and that, "electronic players are focused entirely on the game itself, with very little social small talk, unlike traditional card games."[25] As a result, youth may have less time for civic life, less social capital, and less of the inclination and skills needed for civic engagement.

This perspective, however, is disputed. Some scholars find that Internet use supplements one's social networks by forging additional connections to individuals whom players would not otherwise know, and several have identified mediating variables, such as motivation, that influence the effect of digital engagement.[26] In general, studies of this sort have focused on the Internet broadly (not on video games) and on television. This motivates our interest in the relationship between the quantity of video game play and civic engagement.

Why Study the Civic Characteristics of Video Game Play?

Although game theorists have discussed how the content of video gaming experiences might influence civic outcomes,[27] there has been very little empirical research that examines these relationships. Such research is needed in order to test claims regarding the civic potential of video games and to inform our judgment regarding the likely contribution of particular games and gaming experiences. Moreover, such studies can provide guidance to youth, parents, and educators regarding the desirability of varied games and to game designers who may want to build efficacious features into the games they create.

Although there have been no large-scale quantitative surveys that detail the relationships between the civic characteristics of game play and civic engagement, researchers have identified key features of effective practice in classrooms through controlled, longitudinal, experimental, and quasi-experimental studies in schools and other settings.[28] These features include opportunities to

1. Simulate civic and political activities
2. Voluntarily help others
3. Help guide or direct a given organization or group
4. Learn how governmental, political, economic, and legal systems work
5. Take part in open discussions of ethical, social, and political issues
6. Participate in clubs or organizations where young people have the opportunity to practice productive group norms and to form social networks

These activities are believed to support the development of young people's civic and political commitments, capacities, and connections. In so doing, they are believed to foster development of civic identities while increasing levels of civic activity.

For example, simulations of civic and political activities and learning how government, political, economic, and legal systems work provide young people with the knowledge and skills necessary to participate in the political system.[29]

However, civic participation requires more than knowledge of how institutions work and how people participate in them. It requires an interest in and commitment to participation, which can be developed, for example, through discussions of social issues and through volunteer work to address those issues.[30] It also requires that young people develop confidence in their own abilities (sometimes referred to as a sense of agency) to act as leaders and to work productively for change. To the extent that youth have the opportunity to practice articulating their own point of view, debate issues, and help others in their own communities, they are likely to develop confidence in their ability to do so in the larger civic and political arenas. Finally, civic and political activity is largely a group activity. Youth organizational membership is believed to socialize young people to value and pursue social ties while exposing youth to organizational norms and relevant political and social skills that make maintaining those ties more likely.[31]

The six civic gaming experiences that we attend to in this study closely parallel the six items in this list of "best practices" in civic education.[32] In addition, they align with practices that games researchers have identified as occurring in games. Table 1 describes the characteristics of beneficial in-class curricula and those of civic-based games. The *SimCity* inset reveals some of these characteristics in action. We also describe several video games that provide these civic gaming experiences and discuss research that examines their impact.

Table 1
Best practices for fostering civic responsibility

"Best Practice" Civic Learning Experiences	Examples	Civic Gaming Experiences	Examples
Simulations of civic processes	In Social Studies/ Government class • Simulation of legislative debates • Mock trials	Simulations of civic processes in virtual worlds	*Civilization, SimCity, Rome Total War* • Build new city or civilization • Manage day-to-day operations of city or empire
Instruction in government, history, law, economics, and democracy	Learning about • American Civil War • How a bill becomes a law • Principles of democracy	Games with explicit civic, historical, economic, or legal focus	*The Oregon Trail, Carmen San Diego, Zoo Tycoon, Lemonade Stand* • Games with historical, government, economic content
Community service learning	As part of school unit, volunteer in • VA hospital • Homeless shelter	Service within a gaming community	• Develop game-related Web site with game tips for others • Help "newbies" with game tasks
Extracurricular activities, school club membership	• Participate in school clubs • Write for school newspaper (structured social environments)	Extra–game world activities (formal and informal game communities)	• Join a game guild • Write for a game-related Web site • Participate in chat discussions with other gamers • Research "mods," cheats

| *Student governance and voice* | • Student council
• Student voice in school decisions, e.g. discipline code
• Student voice in classroom decisions | Player governance and voice in game world | *World of Warcraft, EverQuest* (MMOGs)
• Take leadership role in a guild
• Participate in making guild rules and organizational processes
• Build team consensus for goals and strategies for game quest/raid |
| *Discuss/debate/learn about current events and social issues* | Informed discussions about, e.g., immigration, war in Iraq, the economy, in open classroom climate | Discuss and examine current events in games and gaming communities | *Democracy; Decisions, Decisions: Current Issues*
• Games and gaming communities that engage ethical questions
• Games and gaming communities that focus on social problems |

Civic Outcome Goals

People who individually and collectively engage in democratic society in order to identify and address issues of public concern through acts of voluntarism, organizational involvement, and electoral participation.

SimCity is a game with explicit civic content in which players
design and develop their city, considering such aspects as
zoning, land use, taxes, and transportation. Dialogue from an
online community provides a sense of the civic thinking
required by *SimCity* (see box 1).[33]

Box 1

---

sedimenjerry (Traveler) 5/19 2:26 pm
HELP!!! I used to have a large city with a population of about
670,000. Now it is about half of that. Why is the population
decreasing so much? HELP PLEASE

Maxis92 (Dweller) 3:38 pm
Well, your situation is pretty vague and it could be a number of
reasons. Could you give us a brief idea of how your city develop
when it was at 670,000 to now (crime rates, education, jobs,
commute time, pollution, taxes, etc.)

Hahayoudied (Loyalist) 5:59 pm
We can't shoot your problems in the dark, why not give us some
information about your city, and if you have changed it.

sedimenjerry (Traveler) 5/20 1:05 pm
oh sorry that would help. it is on a large city tile and within a
half a year (simcty time) it declined sharply. demand is still high
for commercial res. and industrial. crime has gone down health
is fine garbage has gone down. there are no power or water out-
ages. the only thing i can think of is if the latest NAM and RHW
downloads have affected it. however i have not built any RHW's
in the city. i will try to get a picture of the city

---

i've noticed that the cities are abandoned due to commute time but ive never had this large of a problem. the first pic is the southern region that has the commute problems. the second is of the industrial area and lake city. the third is downtown. i have plenty of subway systems, bus routes, and roads

Maxis92 (Dweller) 6:35 pm
Yeah, well I can only narrow it down to 2 possibilities. You may need to bring more jobs to your city since I'm seeing a lot of "No Job" Zots. That's probably why your demand is high for more commercial jobs. You can do this by placing plenty of plazas and rewards in your business districts. Also, the commute timing will destroy any city, If your sims (especially the wealthy ones) can't find a job only so many minutes from their home, they will quit and probably move elsewhere. Sometimes your subway and bus system may not be efficient and you probably need to fix it or add other alternatives such like an el-trains or a monorails.

[the conversation continues]

sedimenjerry (Traveler) 5/21 12:18
thanks guys its getting larger now

One example of a popular video game with civic content is *Civilization IV*. Players begin with an undeveloped piece of land and a group of settlers. They must make decisions about how to build a city and when to send out scouts to explore surrounding territories, and they must develop warriors to protect the city. Players begin in the Stone Age and move all the way to the

twenty-first century. In the process, they make a range of decisions about when to introduce reading, religion, and the printing press. They negotiate trade agreements and at the same time
are responsible for the day-to-day political and financial governance of the city. Through this simulation, participants have
opportunities to learn about the dynamics of economic, political, and legal systems. Engaging in this way also provides opportunities for participants to develop a civic identity as they see
and experience themselves as civic leaders. Indeed, research in
social psychology finds that such opportunities lead individuals
"to adopt attitudes and cognitions consistent with the behaviors they are acting out." In addition, those engaging in the
simulation have opportunities to practice and develop civic
skills.[34]

A qualitative study by Kurt Squire and Sasha Barab explored
how students used *Civilization III* (the previous version of *Civilization IV*) in a history class to test hypotheses about the influence of such forces as trade, natural resources, and political
alliances on historical events. With guidance and support, students began to appropriate the game for their own educational
(and social) purposes.[35] They developed questions and used the
game to test hypotheses by changing their decision-making
strategies in the game and seeing what then happened.

The Squire and Barab study suggests that young people can
show gains in political and civic knowledge from playing a
commercial video game such as *Civilization*. However, this
occurred in a context where adults guided and shaped the experiences with specific educational goals. It is less clear whether
young people who simply play *Civilization* will have the same
kinds of civic gaming experiences.

In addition to commercially designed games, media researchers have developed games with an explicit educational focus.
For example, *Quest Atlantis*, created as a school-based educational simulation, embeds civic learning opportunities in the
game's play-based educational tasks. Users are youth aged 9 to
12 who participate through their elementary schools or after-
school programs.

Players embark on "Quests" to the fictional world of Atlantis,
which may consist of an online educational activity or be linked
to a real world activity. Atlantis has been taken over by leaders
whose emphasis on progress has contributed to a severe environmental, moral, and social decline. The quests are to help
find solutions to the many problems facing Atlantis. Quests are
aligned with educational standards and a set of social commitments so that students understand the concepts explored in
Quest Atlantis as well as the impact this knowledge has on their
communities. For example, a student might be asked, as part of
the focus on developing a social commitment to environmental
awareness, to identify an animal that lives in the student's area
and to learn about the animal's habitat. The player then would
write a short story based on the information and share it with
the online council of Atlantis.

The game has features that align with "best practice" in civic
education, including simulating civic, political, and economic
processes and researching and discussing personally relevant
social issues. It also provides children with opportunities to discuss the ethical implications of different actions, learn skills
needed to create change around those issues, and connect to
others who are working on the same issue. Moreover, the pro-

cess of playing such games is social and provides opportunities for young people to work collaboratively toward common goals and to express their voice—helping to guide both the strategies that groups of players employ and the way the game itself is played. Finally, *Quest Atlantis* includes a narrative story line using prosocial male and female teen protagonists to help young people understand the purpose behind some of their activities and the interconnections among various activities.

Barab and colleagues have completed several studies that find learning gains from *Quest Atlantis* in science, social studies, and language arts. In social studies, they find significant improvement in students' appreciation for how history relates to their own lives and the ability to adopt multiple perspectives in decision making on international issues.[36]

Although evidence indicates that games can be used productively in an educational setting with some adult intervention and reflection, it is less clear whether gaming in a more typical context, alone or with peers, yields similar benefits. Some argue, however, that with certain design features, games can facilitate powerful civic learning experiences without adult intervention.[37]

Why Study the Social Context of Video Game Play?

Just as prior research by civic educators supports a possible link between certain civic characteristics of video games and civic engagement, the social context of the gaming experience may also be linked to civic engagement. Several well-controlled, longitudinal studies find that adolescents' participation in extracurricular clubs and organizations predicts later civic engagement.[38] This participation is believed to foster social networks and to socialize young people to value and pursue social ties. These experiences also expose members to organizational norms and relevant political and social skills that enable them to maintain these ties.

Thus, if game playing leads to isolation or to integration into gaming communities with antisocial norms, one might expect less civic engagement or connection. On the other hand, to the extent that games are played with others or integrate youth into vibrant communities where healthy group norms are practiced and where teenagers' social networks can develop, games might well develop social capital. Many massively multiplayer online games (MMOGs), for example, do not have explicitly civic or political characteristics, but they require the ongoing and sustained cooperation of a group of people to play. This cooperation can potentially offer teens practice in identifying shared goals, negotiating conflict, and connecting with others who are not part of their daily lives.[39]

Thomas and Seely Brown point out that games such as *World of Warcraft* "involve the experience of acting together to overcome obstacles, managing skills, talents, and relationships and they create contexts in which social awareness, reflection, and joint coordinated action become an essential part of the game experience."[40] Such opportunities can, as Constance Steinkuehler and Dmitri Williams argue, provide a "third place" or form of civil society and civic skill learning.[41] These dynamics lead Jenkins to ask, "who's to say video games are not serving the same purpose that bowling leagues used to provide, where people develop a sense of social responsibility and participation."[42] Some empirical studies have examined these dynamics, but as yet no clear findings have emerged.[43]

Youth have many opportunities to actively engage around the more popular games, including, as Mimi Ito suggests, the creation of "cheats, fan sites, modifications, hacks, walk-throughs, game guides, and various Web sites, blogs, and wikis."[44] These enable players to discuss the game, learn about game options, give tips, and ask for advice. They also provide ways to sidestep the formal constraints of the game and cus-tomize or personalize the gaming experience. Integral to these activities are the opportunities for more experienced players, regardless of age, to take on leadership roles and to help others. The impact of these forms of participation is not yet clear.

Finally, one unique quality of the social nature of game play is that much of it takes place without geographic proximity or face-to-face contact. Although young people can play games together in the same room, new technology makes it possible to play games in highly interactive ways without ever meeting in person. It is unclear whether such online social interaction provides the same opportunities to forge social connections as face-to-face recreational activities.[45] All of these unanswered questions lead us to examine more closely the social context of video game play.

Why Study the Demographic Distribution of Civic Gaming Experiences?

Having identified potentially relevant gaming characteristics and social contexts, we next wanted to assess the prevalence and distribution of such opportunities. At the most basic level, we wanted to understand how common these opportunities are. In addition, it is important to consider the "digital divide" in relation to political participation.[46] Karen Mossberger, Caroline Tolbert, and Ramona McNeal find, for example, that Internet use furthers civic participation but that key kinds of Internet use are unequally distributed and that these inequalities parallel other inequalities in the broader society.[47] We therefore chose to examine whether the digital divide applied to civic gaming experiences. This interest also sprang from our recent findings that white, academically successful children from families with higher education and income have significantly more opportunities for civic learning in school as part of their general curricular and extracurricular activities.[48] In short, we wondered whether the distribution of civic gaming experiences in video games might propagate (or perhaps help redress) the inequalities in civic learning opportunities that exist elsewhere in the society.

Study Design

To explore these questions—whether the frequency of game play, the characteristics of games, and the social context of game play are related to civic engagement and whether gaming experiences that may influence civic engagement are equally distributed—we draw on a phone survey of 1,102 young persons in the United States aged 12 to 17 conducted by the Pew Internet and American Life Project in 2008. The survey recruited teens using random sampling, which allows us to generalize our findings beyond teens who are particularly inclined toward or interested in gaming.

The civic outcomes we monitored were

- searching for information about politics online;
- volunteering in the last 12 months;
- raising money for charity in the last 12 months;
- persuading others how to vote in an election in the last 12 months;
- staying informed about politics or current events during the last 12 months;
- protesting or demonstrating in the last 12 months;

- expressing a commitment to civic participation;
- showing interest in politics.

We examine both interests and activities for several reasons. First, although teens are not politically or civically active in the ways that adults are, both developing commitments and experimenting with engagement are important expressions of young people's emerging civic and political identities. In addition, neither kind of indicator can tell the whole story. Particularly for young adolescents, participation is shaped significantly by parents as well as by their own commitments. On the other hand, if Nie and colleagues are right that digital media detract from time potentially spent on civic issues, then focusing only on interests and commitments will fail to capture the full impact of video game play.

We used statistical methods (multivariate linear and logistic regression) to assess relationships between student background variables and civic gaming experiences, as well as the relationship between quantity and civic quality of gaming and the eight forms of civic engagement noted above. We designated games with the following characteristics to be civic gaming experiences (in contrast to more general experiences):

- helping or guiding other players;
- thinking about moral or ethical issues;
- learning about a problem in society;
- learning about social issues;
- helping to make decisions about how a community, city, or nation should be run;
- organizing or managing game groups or guilds.

Multivariate linear and logistic regressions allow us to control for factors, such as family income or a parent's civic and political activity, that have been shown to influence youth civic participation. Thus we are able to isolate the effects of video gaming on civic engagement above and beyond such factors. For more detail on the methodology, see appendixes A and B.

Measures

To analyze the aspects of game use and their association with civic engagement, we isolated four factors that, given past research, likely influence civic participation, and developed ways to measure that influence. The measures of these factors (listed below), along with measures of civic and political behaviors and attitudes, include the following (see appendix B for a more detailed description):

Demographic Variables   including family income, race, gender, and age

Quantity of Game Play   including items to assess frequency and duration of typical game play

Social Context of Game Play   whether games are played alone, with others in person or with others online, and whether game play is accompanied by secondary social activities

Civic Characteristics of Game Play   whether teens have the civic gaming experiences noted earlier that might promote civic engagement

Civic and Political Behaviors and Attitudes   degree of engagement among teens and their parents in activities ranging from volunteering to participating in elections to protesting, as well as their attitudes about politics and community engagement

Cautionary Note about Causality

Before the discussion of our findings, a caveat is in order. Although this study can identify relationships between civic gaming experiences and civic engagement, it cannot tell us if these experiences directly caused youth to be more or less civically engaged. Experimental and longitudinal data are needed to establish such causal relationships between civic gaming experiences and civic engagement. It may be that gaming experiences promote civic engagement. After all, many civic gaming experiences parallel classroom-based civic learning opportunities that have been shown to foster civic engagement. Yet causality may flow the other way as well. Youth who are more civically inclined and engaged may well seek out games that provide civic gaming experiences. Thus, while analysis of this data can inform the conversation surrounding video games and civic development, more work is needed to fully understand many of the relationships described below.

Findings

The increasing variety and complexity of video games provide young people with a wide range of experiences, including civic gaming experiences. We find that many young people have these experiences, and they have them in a wide range of video games, from strategy games to first-person shooters. We also consider the social contexts in which game play occurs. The findings below describe how the quantity of teens' game play relates to their civic and political engagement. We also examine whether having civic gaming experiences and playing with others (on- and offline) relate to civic outcomes. Finally, we examine how frequently young people are having civic gaming experiences and whether the distribution of these civic gaming experiences is equitable across varied demographic groups.

Research Question 1: The Quantity of Game Play

The quantity of game play is not strongly related (positively or negatively) to most indicators of teens' interest and engagement in civic and political activity.

We compared the civic and political attitudes and behavior of teens who play games at least once a day, those who play games one to five times per week, and those who play games less than once a week or not at all. This investigation is motivated by concerns that children who play a great deal risk becoming socially isolated or experience other negative outcomes. On all eight indicators of civic and political engagement, we find no significant difference, positive or negative, between teens who play every day and those who play less than once a week (after controlling for demographics and parents' civic engagement). That is, those who are the more frequent players are not any less or more likely to engage in social and civic acts than the less frequent players.

On six of the eight indicators, we find no significant differences between teens who play one to five times a week and teens who play less than once a week. The exception is that 11 percent of teens who play games one to five times a week had protested in the past 12 months, compared with 5 percent of teens who play less than once a week. Also, 57 percent of teens who play games one to five times a week say they are interested in politics, compared with 49 percent of teens who play less than once a week. These differences are statistically significant. (See table B.1 in appendix B for details.)

Teens who play every day vary in the number of hours they play each day, ranging from 15 minutes to several hours a day. However, we find only very minor effects of daily time spent playing for two of the eight outcomes. Teens who spend more hours playing games are slightly less likely to volunteer and to express a commitment to civic participation than are those who play for fewer hours (see table B.2 in appendix B for details).

These results suggest that the frequent concerns in the media and elsewhere about the ennui and disconnection among those who play video games for long periods of time may be misplaced.

Research Question 2: The Civic Characteristics of Game Play

The characteristics of teens' gaming experiences are strongly related to their interest and engagement in civic and political activity.

Teens who have civic gaming experiences, such as helping or guiding other players, organizing or managing guilds, playing games that simulate government processes, or playing games that deal with social or moral issues, report much higher levels of civic and political engagement than teens who do not have these kinds of experiences.[49] These differences are statistically significant for seven of the eight civic outcomes we studied (see table B.3 in appendix B for details).[50]

To analyze the relationship between civic gaming experiences and teens' civic and political engagement, we categorize teens into three groups. Those with:

- the fewest civic gaming experiences (in the bottom 25 percent of the distribution of civic gaming experiences);
- average civic gaming experiences (middle 50 percent);
- the most civic gaming experiences (top 25 percent).

Teens with the *fewest* civic gaming experiences may report sometimes helping or guiding other players, but are unlikely to report having any other civic gaming experiences. Teens with *average* civic gaming experiences typically report having several civic gaming experiences at least sometimes or a small number of civic gaming experiences frequently. Teens with the *most* civic gaming experiences typically report having all the civic gaming experiences at least sometimes as well as some civic gaming experiences frequently.

Compared with infrequent gamers, teens who most frequently (top 25 percent) have civic gaming experiences seek out political or current events information. Seventy percent, for example, go online to get information about politics or current events, compared with 55 percent who have infrequent or no civic gaming experiences (see table 2). They also more often raise money for charity, say they are interested in politics, have attempted to persuade someone to vote a particular way, and are more likely to have protested or demonstrated.[51] Those teens who report average amounts (middle 50 percent of users) fall in between frequent and infrequent civic gamers in their levels of civic engagement (see table 2).

Research Question 3: The Social Context of Game Play

Playing games with others in person is related to civic and political engagement.

Teens who play games socially (a majority of teens) are more likely to be civically and politically engaged than teens who play games primarily alone. Among teens who play alongside others in the same room,

- 64 percent have raised money for charity, compared with 55 percent of those who play alone;

Table 2

Teens with more civic gaming experiences are more engaged in civic and political life

Civic and Political Commitments	Teens with Fewest Civic Gaming Experiences (bottom 25%)	Teens with Average Civic Gaming Experiences (middle 50%)	Teens with Most Civic Gaming Experiences (top 25%)
Go online to get information about politics or current events	55	64*	70*
Give or raise money for charity	51	61*	70*
Say they are committed to civic participation	57	61	69*
Say they are interested in politics	41	56*	61*
Stay informed about political issues or current events	49	59*	60*
Volunteer	53	54	55
Persuade others how to vote in an election	17	23	34*
Have participated in a protest march or demonstration	6	7	15*

Source: Pew Internet & American Life Project. Teens, Video Games, and Civics Survey, Nov. 2007–Feb. 2008. Margin of error is ±3%.

* Indicates a statistically significant difference compared to teens with the fewest civic gaming experiences.

• 65 percent go online to get information about politics, compared with 60 percent of those who play alone;

• 64 percent are committed to civic participation, compared with 59 percent of those who play alone;

• 26 percent have tried to persuade others how to vote in an election, compared with 19 percent of those who play alone.

Interestingly, this relationship only holds when teens play alongside others in the same room. Teens who play games with others online are not statistically different in their civic and political engagement from teens who play games alone (see table B.4 in appendix B).

We were curious as to whether the lack of relationship between civic engagement and playing with others online was due to the depth of interactions that occur online. Playing with others online can be a fairly weak form of social interaction, where two players never speak or interact and play only for a short time. It may also include longer and more sustained networks where players join a guild and play games in an ongoing and coordinated fashion. Researchers suggest that the more intensive form of online socializing, for example, in a guild can offer many of the benefits of offline civic spaces that less-intensive online social play may not.[52] To shed light on this issue, we compared those who participate in guilds with those who play alone only. We find no difference between the two groups' level of civic and political engagement. The relationship between guild membership and two civic outcomes (volunteering and raising money for charity) are marginally significant ($p < .10$) (see table B.5 in appendix B). We should point out, however, that organizing and managing game groups or guilds was one of our civic gaming experiences and was associated with greater civic and political engagement.

Youth who socially interact around the game (commenting on Web sites, contributing to discussion boards) are more engaged civically and politically.

Among teens who write or contribute to Web sites or discussion boards related to the games they play, 74 percent are committed to civic participation, compared with 61 percent of those who play games but do not contribute to these online gaming communities. They are also more likely to raise money for charity, stay informed about political events, express interest in politics, try to persuade others to vote in a certain way, and attend protests or demonstrations (see table 3).

These relationships to civic engagement are much weaker among youth who read or visit Web sites, reviews, or discussion boards but who do not write for these sites. We found only one statistically significant difference: among those who visit such sites, 70 percent also go online to get information about politics or current events, compared with 60 percent of teens who play games but do not visit these sites (see table B.6 in appendix B).

Research Question 4: The Demographic Distribution of Civic Gaming Experiences and Social Contexts

Given that the civic characteristics and some of the social contexts of video game play are related to civic engagement, we

Table 3

Teens who contribute to online gaming communities are more engaged in civic and political life than teens who play games but do not contribute to online communities

Civic and Political Commitments	Teens Who Play Games but Do Not Contribute to Game-Related Online Communities	Teens Who Write for or Contribute to Game-Related Online Communities
Say they are committed to civic participation	61	74*
Go online to get information about politics or current events	62	73
Give or raise $ for charity	61	68*
Stay informed about political issues or current events	58	67*
Say they are interested in politics	54	63*
Volunteer	55	58
Persuade others how to vote in an election	22	38*
Have participated in a protest march or demonstration	8	18*

Source: Pew Internet & American Life Project. Teens, Video Games, and Civics Survey, Nov. 2007–Feb. 2008. Margin of error is ±3%.

* Indicates a statistically significant difference compared with teens who play games but do not contribute to online communities.

examine how frequently those who play video games experience these civic characteristics. We also examine how equitably these experiences are distributed.

Many young people have some civic gaming experiences, but few have many.

Between 30 and 76 percent of young people report sometimes experiencing each of the civic gaming experiences listed in table 4. Approximately one-half of teens, for example, have played games that led them to think about moral or ethical issues. However, relatively few teens (typically under 10 percent) report "often" having particular civic gaming experiences.

Different games provide different levels of exposure to civic gaming experiences.

We examine the frequency of the civic gaming experiences among teens who report that one of the five most popular game franchises is one of their three current favorite games. The survey does not enable us to directly assess the civic gaming experiences associated with each game, but a logistic regression that controls for both playing the other popular games and a range of demographic factors provides an estimate of the frequency of civic gaming experiences associated with each game. See table B.7 in appendix B for details of these results.[53]

The five most popular game franchises are *Guitar Hero, Halo, Madden NFL, The Sims,* and *Grand Theft Auto.*[54] We find that playing certain games was associated with more frequent civic gaming experiences:

Table 4

Prevalence of civic gaming experiences

	Teens Who Have the Experience "at Least Sometimes" (%)	Teens Who "Often" Have the Experience (%)
Help or guide other players	76	27
Think about moral or ethical issues	52	13
Learn about a problem in society	44	8
Learn about social issues	40	8
Help make decisions about how a community, city, or nation should be run	43	9
Organize or manage game groups or guilds	30	7

Source: Pew Internet & American Life Project. Teens, Video Games, and Civics Survey, Nov. 2007–Feb. 2008. Margin of error is ±3%. Full question wording: "When you play computer or console games, how often do you ____? Often, sometimes, or never . . . or is that something that does not apply to the games you play?"

- 88 percent of those who report *Halo* as a favorite game report helping or guiding other players, compared with 73 percent of those who do not list *Halo* as a favorite;
- 59 percent of those who list *The Sims* as a favorite game say they learned about problems in society while playing video games, compared with 47 percent who do not list *The Sims* as a favorite;
- 52 percent of those who list *The Sims* as a favorite game say they have explored social issues while playing video games,

compared with 39 percent who do not list *The Sims* as a favorite;

▪ 66 percent of those who list *The Sims* as a favorite say they have made decisions about how a city is run while playing video games, compared with 42 percent who do not list *The Sims* as a favorite.

It is interesting that playing *Halo* is associated with helping or guiding other players. Few likely think of *Halo* as a civically oriented game. *Halo* is a science fiction, first-person shooter game where players must battle to save humankind. That *Halo* players more commonly help and guide other youth speaks to an important observation of new media scholars—that some of the social interactions around certain video games can provide civic gaming experiences.

It is less surprising that *The Sims* franchise provides many civic gaming experiences. *The Sims* is a life simulation game where game play is open-ended. Players create virtual people called "Sims" and then must find housing, look for a job, make decisions about how to spend leisure time, and engage in a wide range of other possible activities. The franchise also includes games such as *SimCity* and *SimTown*, where players engage in explicitly civic activities as they build and guide the development of their own city or town. Each of their decisions has consequences, and players confront multiple dynamics associated with civic and social life. *Sims* is also enormously popular—it is the best-selling PC game in history, with more than 100 million units sold.[55]

Youth play games alone, together with friends, and online with others.

When asked what they do most often, teens are evenly split between solo (49 percent) and group game play (49 percent). Most of the group-gamers play with friends in person, with 77 percent of group-gamers reporting playing games with others in the same room. A small percentage of teens (23 percent) play most often with other people via the Internet. Among those teens who play games with others online, more than two in five (43 percent) say they play games online as a part of group or guild; 54 percent of online gamers are not playing as a part of a group.

Civic gaming experiences and social contexts for game play appear to be equitably distributed by income level, race, and age, although girls have fewer civic gaming experiences.

Interestingly, for civic gaming opportunities, only gender is related to whether teens experience these opportunities. Boys are about twice as likely as girls to report having civic gaming experiences, even when controlling for frequency of game play.[56] Income, race, and age are all unrelated to the amount of reported civic gaming experiences (see table B.8 in appendix B).

Discussion and Implications: The Civic Potential of Video Games

With this first nationally representative quantitative study of the relationship between youth video game play and civic engagement, we hope to inform both scholarly and popular hypotheses about the civic potential of video games. The goal, ultimately, is to better leverage the civic potential of video games.

The findings challenge popular perceptions of gamers as isolated and civically disengaged. They also point to a need for a more nuanced understanding of the ways in which video games relate to civic engagement. For example, we find that the overall amount of game play is unrelated to civic engagement, but that some qualities of game play are strongly related to civic engagement. Likewise, some forms of social activities associated with game play are not related to civic engagement, but others are. Exposure to civic gaming experiences is equitably distributed across most demographic groups, but few youth have frequent civic gaming experiences. We also discuss how parents, educators, policymakers, and advocates might better use games to provide civic gaming experiences. We conclude by outlining avenues for future research on the civic potential of video games.

The study's design limitations somewhat restrict the conclusions we can make about the relationships between civic gaming experiences and civic engagement. Specifically, we could not control for respondents' prior civic commitments, and we did not randomly assign participant exposure to games as would be done in an experimental study. We suspect that the relationships we find between gaming experiences and civic engagement are partially the result of teens with civic interests choosing to play games that provide civic gaming experiences. On the other hand, well-controlled classroom studies (although not with video games) find that these kinds of civic experiences foster civic engagement. More important, even when a young person's civic interest draws him or her to these games, playing such games likely reinforces these interests and further develops civic skills and knowledge.

Given that we cannot make causal claims, our comments should be understood as speculative. Nevertheless, drawing on these findings and on findings from related research, it is possible to offer some preliminary implications that can advance discussions about the civic potential of video games and research priorities.

The stereotype of the antisocial gamer is not reflected in our data. Youth who play games frequently are just as civically and politically active as those who play games infrequently.

Our findings conflict with a commonly held perspective that youth who play video games are socially isolated and often antisocial. We also found no evidence to support scholars' concerns that young people involved in the Internet (in this case by playing video games) are less civically engaged. The quantity of game play, according to our study, is unrelated to most of the civic outcomes measured.

Civic gaming experiences are strongly related to civic engagement.

For those hoping to leverage the civic potential of video games, the strong and consistent relationship between teens' civic gaming experiences and civic engagement is encouraging. It indicates that the same kinds of experiences that foster civic outcomes in well-controlled classroom studies may achieve similar results in gaming environments. Moreover, that the overall quantity of game play is unrelated to civic engagement, but that various qualities of video game play are associated with civic engagement parallels findings from civic education research. The number of civics courses one takes is not strongly related to civic outcomes, yet there is a strong link between civic engagement and particular civic learning opportunities in high-quality civics classes.[57]

These findings are of interest for two main reasons. First, much of the public discussion of video game play frames it in a negative light. These findings show that some gaming experiences are associated with positive civic behaviors. Second, much of the dialogue among new media scholars emphasizes the social aspects of gaming more than the civic content of games. Although we believe the social context of game play, discussed below, is important, this study provides clear reasons to also focus on teens' exposure to civic gaming experiences.

In addition, civic education research leads us to suspect that parents, peers, teachers, and mentors can significantly increase the impact of civic gaming experiences by helping adolescents reflect on those experiences. We draw this parallel from the many studies that have found that the civic value of community service is greatly enhanced when teachers help students reflect on and discuss their experience.[58] This possibility also has implications for game design, which we discuss below.

Social gaming experiences are related to civic engagement in some, but not all, instances.

A core finding from this survey is that gaming is frequently a social activity. Overall, 76 percent of youth play games with others at least some of the time. Youth play with others who are in the room with them and with others online. They organize and manage guilds. They read and contribute to discussion boards. Social interaction in and around many video games is, in other words, common.

It is important to distinguish social interactions that have civic dimensions from those that do not. If four teenagers play basketball together, this activity is social, but not civic. If these four talk with members of their community about the need for lights on a public basketball court, it then becomes a civic activity. A related distinction can be made for online game play and activities. A member of a game guild might focus on developing gaming skills or, alternatively, could be involved in a guild community's decision to prohibit homophobic speech.

In our analysis, the relationship between social participation in and around video games and civic engagement was not consistent. We found that playing games with others in the same room and contributing to Web sites related to a game were associated with civic engagement, but we did not find a statistically significant relationship between playing with others online or as part of a guild and civic engagement.[59] We suspect that part of the reason for these results is related to differing qualities of the social interaction that occurs in these different social contexts. Some of the dimensions of these differences will be discussed below.

The degree to which social life leads to civic engagement in society at large is a matter of much theorizing, empirical study, and debate. Putnam argues that social participation (most famously bowling leagues) can help build a civic culture that supports democracy.[60] Participatory social networks (online social settings where youth interact with their friends and others who share their interests) can lead to participatory civic networks (in which individuals engage with civic and political issues). A variety of factors have been put forward to explain this process. In brief, Putnam and other social theorists argue that social life can foster social capital, which includes trust, social networks, and social norms. Social capital, in turn, is believed to facilitate communication about civic issues, to foster accountability and adherence to desirable social norms, and to enable more effective collective action related to public matters.

Some have argued that particular forms of participation are more likely than others to promote civic engagement. McFarland and Thomas's longitudinal study of extracurricular activities finds that "politically salient youth organizations" (those that involve the kinds of skills and experiences associated with civic and political life), such as student council or a debate club, promote desired civic outcomes. Youth organizations that lack political salience, such as school sports teams, do not.[61] In general, studies of both youth and adults indicate that participation in groups more strongly supports civic outcomes when participants employ civic skills and engage civic topics.[62] This finding is consistent with our research on guild membership. Youth who reported organizing or managing a guild group (a civic skill and one of our civic gaming experiences) were more civically and politically engaged in their offline lives. However, those who were simply members of guilds were not statistically different in their civic and political engagement from those who played games alone.

In addition, our findings and review of the research lead us to suspect that the qualities of a given participatory culture will influence the degree to which it may support a participatory civic culture. For example, we suspect that some guilds create more robust and civically oriented social contexts than others. Civic life requires interactions related to legitimate public concerns.[63] Thus, if interactions are largely about private matters—how to win the game, for example—we would expect them to provide less support for civic life than if the interactions also included broad discussion of current events. Many other factors may matter as well—for example, whether members of online communities also meet face to face to socialize and potentially discuss civic issues; whether participation in an online community is fleeting or long term; whether members of an online community are anonymous; whether norms of civility are modeled and enforced in an online community.

In addition, if the networks developed through video game play are more diverse than the networks youth would otherwise have, and if the social interactions that occur involve more than a narrow focus on the games being played, then they could expand young people's access to different perspectives on many civic or political matters and deepen their general concern for members of society they might otherwise not know. Social gaming experiences might also teach civic skills related to being a member of a group or organizing a group. We suspect that when social interactions teach civic skills or concern civic matters, positive civic outcomes are more likely. As we discuss below, such hypotheses should be a focus of future research.

Civic gaming experiences are more equitably distributed than many other opportunities that support civic engagement.

Given that civic gaming experiences are strongly related to many civic and political outcomes, it is encouraging that they are equitably distributed by race, ethnicity, and family income. The relatively equitable distribution of these civic experiences is important for two reasons. First, this contrasts with teens' experiences in schools and with many forms of Internet use. Specifically, many forms of Internet use that have been found to be related to civic participation are inequitably distributed along lines of race and income.[64] Similarly, in high schools white students and students from higher-income households experience more of the opportunities that support civic and political engagement than do others.[65] For example, students in higher-income school districts are twice as likely as those from average-income districts to learn how laws are made and how

Congress works. They are also more than one-and-one-half times as likely to report having political debates and panel discussions as part of their classroom activities.

Second, civic and political participation among youth is quite unequal. Specifically, much was made of the increasing voting rates of young people in the 2008 primaries, but little mention was made of how unequal this participation was. The voting rate of 18- to 29-year-olds who had attended college was fully three times greater than the voting rates of 18- to 29-year-olds who had not. By equalizing civic learning opportunities, we may be able to help to equalize civic and political participation—a fundamentally important goal in a democracy. Civic gaming experiences may be a means of more equitably developing teens' civic skills and commitments.

It is worth noting that girls experience fewer civic gaming opportunities, even after controlling for the fact that girls play games less frequently than boys. It makes sense to look closely at what may be causing these differences and to consider possible responses.

Few youth have frequent civic gaming experiences.

Although many youth experience some civic gaming experiences, fewer than 10 percent of teens frequently engage in many of the civic gaming experiences we found strongly related to civic outcomes. Increasing the frequency of such experiences is likely necessary to effectively tap the civic potential of video games.

Next Steps for Parents, Educators, and Game Designers

Parents

Parents can increase their children's exposure to civic gaming experiences. As a first step, parents need to be informed about both video games and civic gaming experiences. By being aware of the range of games available and those that specifically offer civic learning experiences, parents can direct their children toward these games. To do this, parents need information both about games with explicit civic content (for example, *Civilization* or *SimCity*) and about what constitutes a civic gaming experience. Organizations such as Common Sense Media might play a role in educating parents by providing civic ratings for games and guides for talking about civic gaming experiences with children. Armed with this information, parents would be able to both make informed choices about which games to purchase and help their children reflect on the civic gaming experiences they have.

Contrary to popular opinion, the games young people play are not all violent. Indeed, as detailed in the Pew report "Teens,

Video Games, and Civics," youth play a wide variety of different video games (we classified 14 different genres), and these games offer a highly varied set of experiences. Just as the desirability of television viewing depends largely on content (watching the History Channel is different from watching cartoons), the desirability of video game play is shaped to a large degree by the content of the experience.

Parents should focus less on the overall quantity of video game play and more on the content and video game experiences their children have, given that we find the quantity of video game play is largely unrelated to civic outcomes, while some qualities of game play are strongly related to civic outcomes. Although there may well be other reasons to limit the quantity of game play (to make more time for physical activity and homework, for example), we suspect that desirability of game play, in many instances, depends heavily on the nature of the game being played.

Parents may be able to guide and influence the games their children play to some extent, yet it is important not to overstate this control. Adolescence, after all, is a time to develop autonomy from parents. Parents may therefore want to work with younger children to help them become thoughtful media consumers (and to develop habits and insights they can carry into their teen years). Parents may also want to play video games with their children. Currently, according to our survey, 31 percent of parents report playing video games with their children at least some of the time. In addition to creating opportunities to have fun together, playing with one's children provides a means to better understand what they are doing and may facilitate valuable conversations about these experiences. Research (largely on TV viewing) suggests that by reflecting with children about their gaming experience, parents can influence how their children think about and interpret the messages and experiences of their game play, including encouraging critical consumption of media.[66] In addition, by discussing political and social issues at home, parents can make their children more aware when they do encounter civic and political content in their video game play. And studies consistently find that youth who discuss civic and political issues with their parents are much more civically engaged than those who do not.[67]

Youth

Teens often make their own decisions about which games they play. Youth perspectives on gaming and on what they find engaging is crucial if compelling games that support civic engagement are to be designed and marketed effectively to youth. In addition, youth often prefer to talk with their friends about the games they play rather than with their parents. It is therefore important to consider how games can be designed to encourage reflection by teens within the game and within the social interactions that surround the game. Given teens' interest in what their peers are thinking and doing, focusing on peer-to-peer learning is particularly important at this age. The game *Zora*, for example, is designed to facilitate this peer-to-peer reflection. It makes discussion of controversial issues an explicit part of the online community and contains a "values dictionary" to foster reflection on and discussion of values, ethics, and rules.[68] Studies assessing the impact of peer-to-peer reflective structures are needed to examine whether such practices amplify the civic impact of playing the game and, if they do, to identify which practices are most effective.

Just as attention to youth perspectives on gaming and on what youth find engaging must be front and center when thinking about ways to better spread the use of games that further the civic potential of video games, it is also very important to attend to youth perceptions of civic and political life. Specifically, many have been asking whether there is a new kind of youth politics. In addition to citing increasing rates of volunteerism, proponents argue that youth participation is often motivated by a different set of concerns than has traditionally been the case—that youth prize action that is informal and grass-roots, and that youth acquire information through alternative means, such as the Internet.[69] Although there can be little doubt that some aspects of youth civic engagement have changed, still up for debate is whether youth civic engagement has been transformed.[70] In either case, both within games and in their offline lives, it is clearly important that youth have space to develop their own ways of engaging civically and, along with such opportunities, that they receive guidance and support from those with more civic and political experience.

Educators

Informing educators about the civic possibilities embedded in some games is another means of increasing the frequency of potentially desirable experiences. Schools and after-school programs provide a direct means of increasing exposure to games that promote civic capacities and commitments. As detailed in "Teens, Video Games, and Civics," one-third of American teens reported playing a computer or console game at school as part of an assignment. The range of games played was broad and included content from math to economic simulations to typing skills. It also included games such as *Oregon Trail* for social studies classes. Although our findings cannot precisely measure the frequency, it is clear that games offering civic gaming experiences can be integrated into the curriculum.

Given this potential, educational organizations and game advocates might reach out to teachers and youth workers, many of whom are unlikely to be aware of ways in which certain video games might support their work. Social studies educators, for example, might be interested in using a game like *Democracy* in a government class. *Democracy* is a multidimensional political simulation in which players respond to varied constituencies, shape policies, and interpret data on approval ratings in an effort to win reelection. Similarly, many global studies educators might be interested in *Real Lives,* in which students can become a different person in a different country. Students then confront decisions, challenges, and opportunities based on the realities of life in those countries. The game can help foster empathy and understanding of the lives of others and teach about dynamics associated with different political systems, economic structures, cultural beliefs, and religions. These games could provide a new and engaging way to teach civics. Indeed, the emphasis on traditional instruction in a civics curriculum has frequently been cited as a major reason civics courses in general have little impact.[71]

Educators can also augment the impact of these experiences and teens' extracurricular gaming experiences by helping young people reflectively engage with video games. Jenkins, for example, highlights a role schools and after-school organizations could play in helping youth develop what he calls "new media literacies." These can support reflection and help youth fully engage in gaming opportunities and problem solve when they run into challenges.[72] Recognizing a related need, teachers implementing *Quest Atlantis* are active participants who guide students through their quest. These teachers receive significant professional development (both online and face-to-face) to effectively monitor student progress and support student reflection and deep thinking in relation to the student's game experience.

Employing a different strategy, at the University of Chicago Charter School Carter G. Woodson Campus, middle-school students are expected to develop the ability to represent their understanding of core academic content through the creation of digital videos, graphics, music (lyrics and instruments), and interactive simulations. For instance, all sixth-grade students are required to learn to create games using *GameStar Mechanic*, a game created to teach students the core principles of game design. Once students have mastered *GameStar Mechanic*, they use their newfound game design skills to create a game that demonstrates their understanding of a scientific concept such as global warming. Noting the potentially important contributions of schools should not, of course, obscure the challenges of integrating desirable forms of video game play into school contexts.[73] We discuss these issues when outlining priorities for research below.

Game Designers

Game designers, in collaboration with civic educators, could create more video games with explicit civic and political content. Such games may well increase the civic impact of video game play. Research on civic education indicates that making explicit connections to civic and political issues is often more efficacious than placing youth in a healthy community context where no explicit connections are made to civic and political issues or skills. A recent study of the development of civic commitments, which controlled for students' prior civic commitments, found that providing students with classroom opportunities to do work explicitly on civic and political issues was more effective than providing supportive school contexts (for example, a caring and supportive school community or a school community where students help one another or work together).[74]

Such findings lead us to suspect that video games that directly engage young people in discussions and collaborative work that explicitly relate to civic or political issues (for example, about the environment, how to govern a city, or how to fight poverty) will be more likely to develop civic skills and commitments. When the focus of the collaboration is not explicitly civic or political (for example, collaborating to solve a puzzle or win a game), we would anticipate less of an impact on civic engagement. Findings from the Pew Teens, Video Games, and Civics Survey are consistent with this. Experiencing frequent civic gaming experiences was strongly related to civic engagement. Playing with others in the same room was only modestly related to civic engagement. Playing with others online or as part of a guild was not significantly related to civic engagement.

Of course, game design is about more than content in the narrow sense. Games that are not explicitly about civics can be designed to develop civic skills and to promote reflective and collaborative dispositions. Flanagan and colleagues note the importance of designing and rewarding prosocial values in both educational and commercial games.[75] In addition, game designers might want to work closely with educators to design games that work more effectively within the structural constraints of many schools and classrooms, while holding onto the core features that make video games so engaging. In addition, game designers might continue to develop strategies for engaging peer-to-peer learning and collaboration in ways that support civic engagement.

Research Agenda

Research That Identifies and Assesses the Impact of Civic Gaming Experience

The widely varying characteristics of teens' gaming experiences highlight the need for research that deepens our understanding of how youth experience video games and how such experiences influence their development (if at all). At this point, most statements regarding the relationships between gaming experiences and civic outcomes are drawn from observations of particular games and gaming dynamics, from correlations between playing games and varied civic indicators, and from what we know from other domains where civic education is practiced. Clearly, these are all worthy places from which to begin considering these issues. However, there is great need for more qualitative and quantitative research that examines teens' video game experiences in relation to civic outcomes.

Ethnographic work in this area will continue to be very important. It can identify, define, and examine features of gaming that have not previously been well conceptualized.

Given the newness and rapidly changing nature of video games, this is particularly important. Ethnographic work can also provide a rich understanding of the significance of context, both the contexts in which youth play games and the ways game play relates to the varied contexts in which youth live. Perhaps most important, ethnographic work enables insight into the ways youth themselves view these experiences, providing an important check on adults' readiness to project particular meanings onto youth.[76]

Consider, for example, a finding from the Pew report "Teens, Video Games, and Civics." The majority of teens, the survey finds, encounter aggressive behavior while playing games. Sixty-three percent reported hearing "people being mean and overly aggressive while playing." Twenty-four percent said this happened often. At the same time, of those who reported having had these experiences, 73 percent said they had heard other players ask the aggressor to stop, with 23 percent reporting that such intervention happens often. Interpreting these responses is difficult. What exactly did youth encounter that they viewed as mean or overly aggressive? What meanings did youth take from these exchanges? Clearly, witnessing the antisocial behavior and responses to it could have civic implications. Witnessing sexist, racist, and homophobic remarks as well as excessively aggressive behavior might heighten a teen's sense of unacceptable behavior. Seeing others intervene might offer productive forms of conflict resolution, skills that will help youth to develop respectful communities.

But none of this is clear. It is difficult to ascertain from survey responses what actually happened during these encounters or to assess how youth experienced these exchanges. If we hope to understand how participation in online communities might shape youth civic commitments and capacities, detailed qualitative inquiry will be necessary to better characterize the range of encounters teens are having and how these encounters are experienced. Such research could then potentially inform the design of further efforts to help youth respond to such episodes more effectively.

Quantitative research in this area will also be very important. The relationships identified in this study between civic gaming experiences and civic engagement, particularly because they align with findings from controlled studies of civic education in other domains, provide an important direction for further inquiry. Currently, however, the lack of controls for young people's prior civic commitments and activities in most existing game research and the lack of random exposure to civic gaming opportunities limit our ability to make causal claims about how games or features of games influence civic development. Longitudinal and experimental studies will enable stronger claims. For example, there is reason to believe that simulations can be designed to foster desired civic outcomes. Studies of how varied simulations influence the development of civic identities and civic skills are needed. Such work provides a way to check the claims of gaming proponents and critics. It can also inform those who do not already have strong opinions about video games, but who are interested in promoting civic goals through video games.

We also found that some types of social experiences around video game play were related to civic engagement, but that others were not. Studying these dynamics with better controls would allow for more nuanced understandings of these dynamics. In addition, crafting questions that more directly get at the different forms of social experiences would allow for deeper insight into differences between the social dynamics of online and face-to-face video game play and into the differences that lead some online video game play to be associated with civic engagement and other such game play to lack this association. For example, does the social or age diversity of groups playing online influence the likelihood that civically oriented issues will arise? Does the relative anonymity of players influence the kinds of norms that are modeled in these communities?

Finally, studies examining the presence of causal relationships between civic gaming experiences and civic engagement should also examine how and why these experiences might bring about shifts in civic engagement. For example, scholars studying civic education have argued that experiences ranging from simulations to learning about and discussing social problems to opportunities to help others can foster a sense of civic capacity (or agency), commitment to particular issues, and connection to others who hold similar concerns. These capacities, commitments, and connections are the building blocks of a civic identity.[77] Other related perspectives and questions are worth considering as well. For instance, do certain games allow more agency, imagination, or creativity in game play around civic issues than others? Does this greater sense of agency affect levels of civic engagement? Deepening our understanding about why playing certain video games furthers civic engagement might well help both educators and game designers better maximize the civic potential of some gaming experiences.

Research on the Role Schools Can Play

The focus on intentional efforts discussed above highlights a key question for research and policy: Can and will schools effectively support the delivery of civic gaming experiences?

There is understandable hesitancy on the part of many proponents of digital media to engage with schools. Schools often fail to deliver the kind of active, student-directed learning that the best video games model. Nationally, for example, 90 percent of ninth graders said reading textbooks and doing worksheets was their most common activity in social studies.[78]

The factors that enable and constrain effective use of video games in schools need to be studied. Such studies might chronicle more- and less-effective efforts to confront the challenges reformers face, ranging from aligning game content with academic standards, to technical challenges associated with using computers in classrooms, to ways to help educators appreciate the potential that some video games represent, to costs associated with the hardware and software that games require.[79]

It is also important to study which students are given these opportunities. As noted earlier, students who are white, from families with higher incomes, or more academically able often have access to many more civic learning opportunities in school than do other students. If video gaming in schools follows this pattern, the use of games may exacerbate political inequality. On the other hand, if games are provided to a broad cross section of students, they might help to lessen inequalities in civic education. In this study, we found that students of varying income, race, and age all report similar levels of civic gaming experiences. To the extent that schools provide such experiences, it would be important to know whether they do so equally as well.

Finally, it is clearly important to study the impact of video games when used in schools. A helpful fact about doing such studies in schools is that students are often randomly assigned to classrooms, which makes it easier to approximate experimental conditions. When undertaking such studies, it will be very important to identify appropriate outcomes and related indicators. Games designed to promote civic skills and commitments may not be well suited to boost math test scores. Unfortunately, the pressure to influence standard academic outcomes often leads educational interventions to be assessed on outcomes that do not align with the intervention's goals.

Research on Civic and Democratic Decision Making

The Teens, Video Games, and Civics Survey focused on civic engagement. Clearly, in addition to levels of engagement, democratic societies must be concerned with the knowledge, analysis, and goals that inform those actions. Assessing such efforts might require, for example, gauging teens' critical analysis, attention to accurate information, and consideration of alternative perspectives. Games might well promote these outcomes. For example, games can place people in a variety of roles. In doing so, they may be able to help players consider alternative perspectives. Similarly, games might well be effective ways to foster civic knowledge, strategic thinking, and consideration of differing stances with respect to pressing social issues. Studies that examine how different games do (or do not) effectively respond to such goals would be valuable.

Research on Other Pathways to Participation

The Teens, Video Games, and Civics Survey was designed to assess the degree to which video games promote the kinds of civic learning opportunities that civic educators associate with best practice. Other dynamics associated with playing video games may also relate to civic outcomes. For instance, many have stressed the importance of recruitment into political activities as a main way for youth and adults to become engaged.[80] The social networks young people develop through gaming (and those they may abandon due to the time demands of gaming) may make recruitment more or less likely. Or it may be that certain forms of video game play make recruitment into some kinds of civic and political life more likely than recruitment into other kinds of civic and political life. Studies assessing such possibilities would be valuable.[81]

Research on Video Games and the Development of Democratic (or Anti-Democratic) Values

Some games have been criticized for promoting masculine values and stereotypes of women and persons of color.[82] It is important to assess such possibilities and also their reverse. Can games designed to challenge problematic stereotypes have their desired effect? Similarly, some games may influence how teens think about social issues such as poverty, war, their environment, or gang life. Games may also influence players' perspectives on possible responses to varied social problems. Finally, some scholars are considering how video games might influence young people's developing perspectives on democratic citizenship.[83] Chad Raphael, Christine Bachen, and colleagues, for example, have put forward a framework that generates hypotheses about how design features (such as the way ethical judgments are incorporated into games) can influence the development of democratic values.[84] Testing the hypotheses embedded in such frameworks will deepen our understanding of the differing kinds of democratic values video games may promote. Developing a better understanding of how the content and structure of games influence such outcomes is important if we wish to fully tap the civic potential of video games.

Conclusion

Judged by any standard, video games are enormously popular. If, in the past, video games were considered a supplement to such media mainstays as television and the movies, this is no longer the case. The April 2008 video game release of *Grand Theft Auto IV* grossed a staggering $310 million in sales on its first day.[85] This was twice the largest domestic movie premiere to date.[86] Not only are these games popular, but they are often deeply engaging and, as a result, may well influence a wide range of attitudes and behaviors. Studying the nature of this influence is therefore of great importance, so that we can better understand and help guide engagement with this powerful force in youth culture.

Appendix A: Parent and Teen Survey on Gaming and Civic Engagement Methodology

Prepared by Princeton Survey Research Associates International for the Pew Internet and American Life Project

Summary

The Teens, Video Games, and Civics Survey, sponsored by the Pew Internet and American Life Project, obtained telephone interviews with a nationally representative sample of 1,102 12- to 17-year-olds and their parents in continental U.S. telephone households. The survey was conducted by Princeton Survey Research International. Interviews were done in English by Princeton Data Source, LLC from November 1, 2007 to February 5, 2008. Statistical results are weighted to correct known demographic discrepancies. The margin of sampling error for the complete set of weighted data is ±3.2 percent. Details on the design, execution, and analysis of the survey are discussed below.

Design and Data Collection Procedures

Sample Design

The sample was designed to represent all teens ages 12 to 17 living in continental U.S. telephone households. The telephone sample was provided by Survey Sampling International, LLC (SSI) according to PSRAI specifications. The sample was drawn using standard *list-assisted random digit dialing* (RDD) methodology. *Active blocks* of telephone numbers (area code + exchange + two-digit block number) that contained three or more residential directory listings were selected with probabilities in proportion to their share of listed telephone households; after selection two more digits were added randomly to complete the number. This method guarantees coverage of every assigned phone number regardless of whether that number is directory listed, purposely unlisted, or too new to be listed. After selection, the numbers were compared against business directories and matching numbers purged.

Contact Procedures

Interviews were conducted from November 1, 2007, to February 5, 2008. As many as 10 attempts were made to contact every sampled telephone number. Sample was released for interviewing in replicates, which are representative subsamples of the larger sample. Using replicates to control the release of sample ensures that complete call procedures are followed for the entire sample. Calls were staggered over times of day and days of the week to maximize the chance of making contact with potential respondents. Each household received at least one daytime call in an attempt to find someone at home. In each contacted household, interviewers first determined if a child aged 12 to 17 lived in the household. Households with no children in the target age range were screened out as ineligible. For eligible households, interviewers first conducted a short interview with a parent or guardian and then interviews were conducted with the target child.[87]

Weighting and Analysis

Weighting is generally used in survey analysis to compensate for patterns of nonresponse that might bias results. The interviewed sample of all adults was weighted to match national parameters for both parent and child demographics. The parent demographics used for weighting were sex, age, education, race, Hispanic origin, and region (U.S. Census definitions). The child demographics used for weighting were gender and age. These parameters came from a special analysis of the Census Bureau's 2006 Annual Social and Economic Supplement (ASEC) that included all households in the continental United States that had a telephone.

Weighting was accomplished using Sample Balancing, a special iterative sample weighting program that simultaneously balances the distributions of all variables using a statistical technique called the Deming Algorithm. Weights were trimmed to prevent individual interviews from having too much influence on the final results. The use of these weights in statistical analysis ensures that the demographic characteristics of the sample closely approximate the demographic characteristics of the national population. Table A.1 compares weighted and unweighted sample distributions to population parameters.

Effects of Sample Design on Statistical Inference

Post–data collection statistical adjustments require analysis procedures that reflect departures from simple random sampling. PSRAI calculates the effects of these design features so that an appropriate adjustment can be incorporated into tests of statistical significance when using these data. The so-called "design effect" or *deff* represents the loss in statistical efficiency that results from systematic nonresponse. The total sample design effect for this survey is 1.17.

PSRAI calculates the composite design effect for a sample of size n, with each case having a weight w_i, as

$$deff = \frac{n \sum\limits_{i=1}^{n} w_i^2}{\left(\sum\limits_{i=1}^{n} w_i\right)^2}$$

In a wide range of situations, the adjusted *standard error* of a statistic should be calculated by multiplying the usual formula by the square root of the design effect ($\sqrt{deff}$). Thus, the formula for computing the 95 percent confidence interval around a percentage is

$$\hat{p} \pm \left(\sqrt{deff} \times 1.96 \sqrt{\frac{\hat{p}(1-\hat{p})}{n}} \right)$$

Table A.1

Sample demographics

	2006 Parameter	Unweighted	Weighted
Census Region			
Northeast	18.2	17.5	18.2
Midwest	22.3	27.0	22.9
South	35.6	33.1	35.5
West	23.9	22.3	23.3
Parent's Sex			
Male	44.1	36.7	43.2
Female	55.9	63.3	56.8
Parent's Age			
< 35	10.0	8.0	9.6
35–39	19.0	16.2	18.8
40–44	28.4	24.7	28.2
45–49	24.4	26.7	24.7
50–54	12.4	15.3	12.6
55+	5.8	9.1	6.2
Parent's Education			
Less than HS grad.	12.6	6.6	10.9
HS grad.	35.5	28.0	35.8
Some college	22.9	26.4	23.3
College grad.	29.0	39.0	30.0
Parent's Race/Ethnicity			
White/Hispanic	66.3	74.6	68.0
Black/Hispanic	11.4	11.1	11.6
Hispanic	16.3	9.5	14.4
Other/Hispanic	6.0	4.8	6.0
Child's Sex			
Male	51.2	50.5	51.1
Female	48.8	49.5	48.9
Child's Age			
12	16.7	14.7	16.5
13	16.7	16.5	16.7
14	16.7	14.2	16.4
15	16.7	18.4	17.0
16	16.7	17.9	16.7
17	16.7	18.3	16.8

where $\hat{p}$ is the sample estimate and n is the unweighted number of sample cases in the group being considered.

The survey's *margin of error* is the largest 95 percent confidence interval for any estimated proportion based on the total sample—the one around 50 percent. For example, the margin of error for the entire sample is ±3.2 percent. This means that in 95 out every 100 samples drawn using the same methodology, estimated proportions based on the entire sample will be no more than 3.2 percentage points away from their true values in the population. The margin of error for teen Internet users is ±3.3 percent and for teen game players is ±3.2 percent. It is important to remember that sampling fluctuations are only one possible source of error in a survey estimate. Other sources, such as respondent selection bias, questionnaire wording and reporting inaccuracy, may contribute additional error of greater or lesser magnitude.

Response Rate

Table A.2 reports the disposition of all sampled telephone numbers ever dialed from the original telephone number sample. The response rate estimates the fraction of all eligible respondents in the sample that were ultimately interviewed. At PSRAI it is calculated by taking the product of three component rates[88]:

- contact rate (the proportion of working numbers where a request for interview was made) of 84 percent[89]
- cooperation rate (the proportion of contacted numbers where a consent for interview was at least initially obtained, versus those refused) of 41 percent

• completion rate (the proportion of initially cooperating and eligible interviews that were completed) of 78 percent

Thus the response rate for this survey was 26 percent.

Table A.2
Sample disposition

112,882	*Total numbers dialed*
6,768	*Business/government/nonresidential*
5,949	*Fax/modem*
62	*Cell phone*
42,092	*Other not-working*
8,181	*Additional projected NW*
49,830	Working numbers
44.1%	***Working rate***
2,430	*No answer*
298	*Busy*
4,677	*Answering machine*
731	*Other non-contacts*
41,695	Contacted numbers
83.7%	***Contact rate***
2,244	*Callbacks*
22,567	*Refusal 1—refusal before eligibility status known*
16,884	Cooperating numbers
40.5%	***Cooperation rate***
1,824	*Language barrier*
13,647	*Screenouts*
1,413	Eligible numbers
8.4%	***Eligibility rate***
311	*Refusal 2—refusal after case determined eligible*
1,102	Completes
78.0%	***Completion rate***
26.4%	***Response rate***

Appendix B: Regression Analysis

The findings regarding the relationships among frequency, social context, and civic qualities of gaming experiences and civic engagement were derived using regression analysis. This statistical technique allows us to pinpoint whether a relationship between different gaming experiences and civic and political engagement exists after controlling for factors such as income, race, gender, and parent involvement—all individual characteristics that have been previously found to be important predictors of civic and political engagement.

Logistic regression was used in conducting the analyses, with the dependent variables being:

- Go online to get information about politics (Yes/No);
- Volunteered in the last 12 months (Yes/No);
- Raised money for charity in the last 12 months (Yes/No);
- Persuaded others how to vote in an election in the last 12 months (Yes/No);
- Stayed informed about politics or current events during the last 12 months (Yes/No);
- Protested in the last 12 months (Yes/No);
- Commitment to civic participation (Agree/Disagree);

- Interest in politics (Agree/Disagree).

To determine the relationship between frequency of gaming experiences and civic and political engagement, each of the outcomes was modeled as a function of the following variables:

Demographic Parent income[90] (a scale that runs from 1 to 8), race (white, African American, Hispanic, or other), gender, and age (binary variable with two categories: 12–14, 15–17).

Parent Involvement Included parent reports of whether, in the last 12 months, they volunteered, raised money for charity, protested, or stayed informed about politics or current events. For each outcome, the parental involvement item that most closely matched the outcome was included in the analysis.

Frequency of Game Play Frequency of game play was measured on an ordinal scale from 1 to 6, ranging from less than once a week to several times a day. For this analysis, frequency of game play was transformed into three categories—(1) every few weeks or less, (2) one to five days a week, (3) daily or more. In all regression models, frequency of game play was entered as a dummy variable with the lowest frequency serving as the reference group.

To determine the relationship between the social context of game play and civic and political engagement, each outcome was modeled as a function of the demographic and parent involvement variables described above and

Playing Games with Others in Person For the game he or she plays most often, teen played games with other people who were in the same room as them (Yes/No);

Playing Games with Others Online   For the game they play most often, teen played the game with people who were connected to them through the Internet (Yes/No);

Researching the Game   Teen read or visited Web sites, reviews, or discussion boards related to the games they play (Yes/No);

Contributing to Online Writing or Discussion about the Game Teen wrote for or contributed to Web sites, reviews, or discussion boards related to the games they play (Yes/No).

To determine the relationship between civic gaming experiences and civic and political engagement, each outcome was modeled as a function of the demographic and parent involvement variables described above and

Civic Gaming Experiences   The civic gaming experiences variable was created by averaging six items measured on a three-point scale (never, sometimes, often). This continuous variable was then broken into three categorical variables—fewest civic gaming experiences, average civic gaming experiences, and most civic gaming experiences. Most civic gaming experiences included teens in the top 25 percent of frequency, average civic gaming experiences included teens in the middle 50 percent, and fewest gaming experiences included teens who fell into the bottom 25 percent. In all regressions, the variable was entered as a dummy variable with infrequent civic gaming experiences serving as the reference group.

Finally, distribution of civic gaming experiences was analyzed using binary logistic regression with civic gaming experiences as the outcome (Infrequent vs. Average or Frequent), modeled as a function of demographic variables (parent income, race, gender, age) and frequency of game play, which are described above.

Table B.1

Relationship between frequency of game play and civic and political engagement

	Civic and Political Outcomes					
	Get Info about Politics	Volunteer	Charity	Stay Informed	Protest	Political Interest
	Exp(B)	Exp(B)	Exp(B)	Exp(B)	Exp(B)	Exp(B)
Demographic Variables						
Income	1.062	1.084	0.977	1.104*	0.920	1.080
Parent Hispanic	1.631	.619*	0.892	0.835	1.366	0.771
Parent African American	1.152	0.682	0.802	1.117	1.120	1.208
Parent Other	2.582*	1.630	1.010	1.990*	0.529	1.207
Child age (older)	1.426*	1.361*	1.091	1.982***	1.027	1.705***
Child sex (female)	1.013	1.213	1.305	1.180	1.585	1.090
Parent Involvement						
Parent volunteered	—	2.208***	—	—	—	—
Parent charity	—	—	2.047***	—	—	—
Parent protested	—	—	—	—	4.901***	2.277*
Parent stays informed	1.156	—	—	2.575***	—	0.935
Frequency of Game Play						
Some games (vs. little/none)	1.044	0.982	1.187	1.078	2.545*	1.684**
Frequent games (vs. little/none)	0.677	0.698	0.939	0.781	1.878	1.265
R²	.046**	.102***	.051***	.119***	.065**	.054**

Notes: For two of the civic and political outcomes measured, persuading others how to vote in an election and commitment to civic participation, the omnibus test was nonsignificant. Those outcomes are excluded from the table.

* *p* < .05; ** *p* < .01; *** *p* < .001

Table B.2

Relationship between hours of game play and civic and political engagement

	Civic and Political Outcomes					
	Get Info about Politics	Volunteer	Charity	Stay Informed	Commitment to Participation	Political Interest
	Exp(B)	Exp(B)	Exp(B)	Exp(B)	Exp(B)	Exp(B)
Demographic Variables						
Income	1.063	1.080	0.972	1.099*	.882*	1.080
Parent Hispanic	1.666*	.630*	0.905	0.856	0.885	0.769
Parent African American	1.153	0.701	0.807	1.135	0.813	1.205
Parent Other	2.725*	1.679	1.025	2.071*	1.278	1.217
Child age (older)	1.465*	1.380*	1.071	2.002***	1.384	1.682**
Child sex (female)	1.017	1.182	1.187	1.150	1.174	0.956
Parent Involvement						
Parent volunteered	—	2.232***	—	—	1.441*	—
Parent charity	—	—	2.089***	—	1.234	—
Parent protested	—	—	—	—	0.966	2.199*
Parent stays informed	1.165	—	—	2.584***	1.321	0.932
Hours of Game Play						
Hours of game play	0.929	.855*	0.881	0.917	.829*	0.954
R^2	.036*	.103***	.053***	.115***	.054**	.041*

Notes: For two of the civic and political outcomes measured, persuading others how to vote in an election and protesting, the omnibus test was nonsignificant. Those outcomes are excluded from the table.

* $p < .05$; ** $p < .01$; *** $p < .001$

Table B.3

Relationship between civic gaming experiences and civic and political engagement

	Get Info about Politics	Volunteer	Charity	Persuade Others	Stay Informed	Protest	Participatory Citizenship	Political Interest
	Civic and Political Outcomes							
	Exp(B)	Exp(B)	Exp(B)	Exp(B)	Exp(B)	Exp(B)	Exp(B)	Exp(B)
Demographic Variables								
Income	1.093	1.097*	0.998	1.049	1.113*	0.933	.902*	1.108*
Parent Hispanic	1.619	.624*	0.886	0.946	0.853	1.313	0.857	0.748
Parent African American	1.149	0.692	0.793	1.096	1.053	1.189	0.776	1.131
Parent Other	2.679*	1.659	0.974	0.945	1.942	0.528	1.259	1.138
Child age (older)	1.570**	1.417*	1.171	1.516*	2.115***	1.146	1.471*	1.742**
Child sex (female)	1.124	1.331	1.422*	1.356	1.289	1.394	1.329	1.079
Parent Involvement								
Parent volunteered	—	2.271***	—	—	—	—	1.510*	—
Parent charity	—	—	2.094***	—	—	—	1.177	—
Parent protested	—	—	—	—	—	5.139***	0.977	2.345*
Parent stays informed	1.136	—	—	1.269	2.588***	—	1.268	0.897
Civic Gaming Experiences								
Average civic gaming Experiences	1.635*	1.310	2.175***	1.586	2.241***	2.060	1.394	2.092***
Most civic gaming Experiences	2.624***	1.461	3.095***	3.327***	1.976**	3.307**	2.024**	2.657***
R^2	.066***	.101***	.092***	.064***	.141***	.075**	.059**	.077***

Notes: * $p<.05$; ** $p<.01$; *** $p<.001$

Table B.4

Relationship between playing with others and civic and political engagement

	Civic and Political Outcomes							
	Get Info about Politics	Volunteer	Charity	Persuade Others	Stay Informed	Protest	Participatory Citizenship	Political Interest
	Exp(B)	Exp(B)	Exp(B)	Exp(B)	Exp(B)	Exp(B)	Exp(B)	Exp(B)
Demographic Variables								
Income	1.080	1.092*	0.985	1.032	1.112*	0.937	.896*	1.090
Parent Hispanic	1.645*	0.637	0.890	1.021	0.845	1.361	0.887	0.761
Parent African American	1.093	0.679	0.801	1.050	1.101	1.068	0.789	1.170
Parent Other	2.638**	1.608	1.009	0.979	2.000*	0.519	1.298	1.185
Child age (older)	1.505***	1.406*	1.100	1.454*	2.023***	1.002	1.398*	1.702***
Child sex (female)	1.099	1.339**	1.300	1.330	1.237	1.369	1.269	1.005
Parent Involvement								
Parent volunteered	—	2.296***	—	—	—	—	1.482*	—
Parent charity	—	—	2.139***	—	—	—	1.203	—
Parent protested	—	—	—	—	—	4.240***	0.930	2.130
Parent stays informed	1.160	—	—	1.331	2.594***	—	1.321	0.936
Social Context								
Play games with others in person	1.397*	1.138	1.662**	1.738**	0.960	1.070	1.424*	1.147
Play games with others online	1.211	1.325	1.008	1.446	1.234	1.334	0.940	1.183
R^2	.043**	.099***	.065***	.040*	.116***	0.049	.052**	.044**

Notes: * $p < .05$; ** $p < .01$; *** $p < .001$

Table B.5

Relationship between guild membership and civic and political outcomes

	Civic and Political Outcomes							
	Get Info about Politics	Volunteer	Charity	Persuade Others	Stay Informed	Protest	Participatory Citizenship	Political Interest
	Exp(B)	Exp(B)	Exp(B)	Exp(B)	Exp(B)	Exp(B)	Exp(B)	Exp(B)
Demographic Variables								
Income	1.062	1.080	0.975	0.944	1.121	1.029	0.942	1.072
Parent Hispanic	1.662	0.590	1.102	1.022	1.098	0.995	0.498	0.842
Parent African American	1.706	1.101	1.205	0.813	0.724	0.000	1.156	0.929
Parent Other	2.850	1.831	1.877	1.171	2.067	0.000	1.170	1.197
Child age (older)	1.288	0.992	1.083	1.789†	1.347	0.763	1.005	1.638†
Child sex (female)	1.133	1.527	0.975	1.361	0.852	1.236	0.997	1.033
Parent Involvement								
Parent volunteered	—	1.845*	—	—	—	—	1.425	—
Parent charity	—	—	2.609**	—	—	—	0.958	—
Parent protested	—	—	—	—	—	2.711	0.935	3.768†
Parent stays informed	0.855	—	—	0.938	3.661***	—	1.120	1.264
Social Context								
Play games in guild vs. Play alone only	1.030	1.663†	1.667†	1.408	0.906	1.803	1.115	1.288
R^2	0.030	.080*	.080*	0.040	.121**	0.104	0.036	0.057

Notes: * $p<.05$; ** $p<.01$; *** $p<.001$; † $p<.10$

Table B.6

Relationship between researching and writing about games and civic and political engagement (controlling for social nature of game play)

	Civic and Political Outcomes							
	Get Info about Politics	Volunteer	Charity	Persuade Others	Stay Informed	Protest	Participatory Citizenship	Political Interest
	Exp(B)	Exp(B)	Exp(B)	Exp(B)	Exp(B)	Exp(B)	Exp(B)	Exp(B)
Demographic Variables								
Income	1.097*	1.082	1.016	1.093	1.116*	0.937	.900*	1.093
Parent Hispanic	1.593	0.711	1.044	1.056	0.859	1.454	0.976	0.670
Parent African American	1.158	0.690	0.710	0.929	1.162	1.458	0.692	1.227
Parent Other	2.787**	1.673	1.229	0.995	2.435*	0.554	1.432	1.284
Child age (older)	1.566**	1.421*	1.065	1.546*	1.956***	1.008	1.424*	1.654**
Child sex (female)	1.255	1.211	1.223	1.388	1.276	1.417	1.167	0.990
Parent Involvement								
Parent volunteered	—	2.281***	—	—	—	—	1.529*	—
Parent charity	—	—	2.102***	—	—	—	1.142	—
Parent protested	—	—	—	—	—	4.748***	1.045	2.322*
Parent stays informed	1.130	—	—	1.507	2.667***	—	1.393	1.115

Social Context								
Play games with others in person	1.355	1.031	1.564**	1.804**	0.897	0.891	1.446*	1.115
Play games with others online	1.088	1.060	0.910	1.181	0.920	0.748	0.779	0.986
Research game play	1.716**	0.876	0.864	1.173	1.145	1.517	1.149	1.079
Write about game play	1.132	1.585	1.892*	2.667***	1.835*	2.870**	1.881*	1.738*
R^2	.061***	.092***	.071***	.804***	.118***	.096**	.071**	.048*

Notes:

* $p < .05$

** $p < .01$

*** $p < .001$

Table B.7

Relationship between participants' most-played game franchises and civic gaming experiences

	Civic Gaming Experiences				
	Helped or Guided other Players	Learned about Problems in Society	Explored Social Issues	Made Decisions How City etc. Is Run	Organized or Managed Guilds or Groups
	Exp(B)	Exp(B)	Exp(B)	Exp(B)	Exp(B)
Demographic Variables					
Income	0.999	.901*	.853**	1.078	.903*
Hispanic	0.863	0.834	1.167	0.687	1.439
African American	1.224	0.808	1.309	0.824	0.771
Other	1.140	1.755	0.660	0.704	1.106
Age	.643*	0.961	0.784	0.894	0.896
Gender	.464**	.476***	0.937	0.775	0.702
Game Franchises					
GTA	0.643	1.026	0.855	1.587	0.759
Sims	2.182	**3.444*****	**2.193****	**3.432*****	1.148
Halo	**2.582****	1.257	1.376	1.132	1.457
Guitar Hero	1.614	0.734	0.939	0.687	0.884
Madden	0.800	0.696	1.234	0.870	1.238
R^2	.099***	.079***	.060**	.057***	.042*

Notes: For one of the civic gaming experiences, "thinking about moral or ethical issues," the omnibus test was nonsignificant. This outcome was excluded from the table.

* $p < .05$; ** $p < .01$; *** $p < .001$

Table B.8
Demographic predictors of civic gaming experiences

	Reported Having "Some" or "Frequent" Civic Gaming Experiences
	Exp(B)
Demographic Variables	
Income	0.930
Parent Hispanic	0.922
Parent African American	1.370
Parent Other	1.443
Child age (older)	0.844
Child sex (female)	.667*
Frequency of Game Play	
Some games (vs. little/none)	1.595*
Frequent games (vs. little/none)	1.936**
R^2	.048**

Notes:

* $p < .05$

** $p < .01$

*** $p < .001$

Notes

1. Amanda Lenhart et al., "Teens, Video Games, and Civics," *Pew Internet and American Life Report*, September 16, 2008.

2. Ibid.

3. Mizuko Ito et al., *Hanging Out, Messing Around, and Geeking Out: Living and Learning with New Media* (Cambridge, MA: MIT Press, forthcoming).

4. Cheats are cheat codes that make changes to the way a video game works. They might give a player new abilities, for example. Mods are modifications to a video game. These might involve new content, characters, or music, for example.

5. National Institute on Media and the Family, "Fact Sheet: Effects of Video Game Playing on Children" (Minneapolis: National Institute on Media and the Family, 2008). http://www.mediafamily.org/facts/factseffect.shtml (accessed July 8, 2008).

6. David Shaffer et al., "Video Games and the Future of Learning," *Phi Delta Kappan* 87 (2005): 104–111; Constance Steinkuehler, "Learning in Massively Multiplayer Online Games," in "Cognition and Learning in Massively Multiplayer Online Games: A Critical Approach" (PhD dissertation, University of Wisconsin–Madison, 2005); James Paul Gee, *What Videogames Can Teach Us about Learning and Literacy* (New York: Pal-

grave/MacMillan, 2003); Katie Salen, "Gaming Literacies: What Kids Learn Through Design," *Journal of Educational Multimedia and Hypermedia* 16, no. 3 (2003): 301–322; Eric Klopfer, *Augmented Learning: Research and Design of Mobile Educational Games* (Cambridge, MA: MIT Press, 2008).

7. Henry Jenkins, "Confronting the Challenges of Participatory Culture: Media Education for the Twenty-First Century" (white paper, MacArthur Foundation Digital Media and Learning Program, Chicago, 2006).

8. Shaffer et al., "Video Games and the Future of Learning"; see also Craig A. Anderson, Douglas A. Gentile, and Katherine E. Buckley, *Violent Video Game Effects on Children and Adolescents: Theory, Research, and Public Policy* (Oxford: Oxford University Press, 2007).

9. Jenkins, "Confronting the Challenges."

10. Norman Nie and Lutz Erbring, "Internet and Society: A Preliminary Report," *IT and Society* 1 (2002): 275–283.

11. John Dewey, *Democracy and Education* (New York: Free Press, 1916), 83. Also see John Dewey, *The Public and Its Problems* (Athens, OH: Swallow Press, 1927 [1954]).

12. Benjamin Barber, *Strong Democracy: Participatory Politics for a New Age* (Berkeley: University of California Press, 1984); Harry Chatten Boyte and Nancy N. Kari, *Building America: The Democratic Promise of Public Work* (Philadelphia: Temple University Press, 1996).

13. CIRCLE, "Research: 2006 Civic and Political Health of the Nation Survey" (New York: Center for Information and Research on Civic Learning and Engagement, 2006). http://www.civicyouth.org/research/products/youth_index_2006.htm (accessed June 13, 2008).

14. Anthony Lutkus and Andrew R. Weiss, *The Nation's Report Card: Civics 2006* (National Center for Education Statistics NCES no. 2007-476) (Washington, DC: U.S. Government Printing Office, 2007).

Recently, there have been some encouraging indications that some aspects of youth civic participation are improving. Most notably, the voting rates of those under age 30 have improved markedly since 2000. See Emily Hoban Kirby et al., *The Youth Vote in the 2008 Primaries and Caucuses* (New York: Center for Information and Research on Civic Learning and Engagement, June 2008).

15. Stephen Macedo et al., *Democracy at Risk: How Political Choices Undermine Citizen Participation, and What We Can Do about It* (Washington, DC: Brookings Institution, 2005), 1.

16. Miranda Yates and James Youniss, "Community Service and Political Identity Development in Adolescence," *Journal of Social Issues* 54, no. 3 (1998): 495–512; Robert Atkins and Daniel Hart, "Neighborhoods, Adults, and the Development of Civic Identity in Urban Youth," *Applied Developmental Science* 7, no. 3 (2003): 156–164.

17. Erik Erikson, *Identity: Youth and Crisis* (New York: W. W. Norton, 1968).

18. Cynthia Gibson and Peter Levine, *The Civic Mission of Schools* (New York: The Carnegie Corporation of New York and the Center for Information and Research on Civic Learning, 2003).

19. Jenkins, "Confronting the Challenges."

20. Ibid., 6.

21. Doug Thomas and John Seely Brown, "Why Virtual Worlds Can Matter" (working paper, University of Southern California, Institute for Network Culture, 2007), 15. http://www.johnseelybrown.com/needvirtualworlds.pdf (accessed May 12, 2008).

22. Constance A. Steinkuehler, "The New Third Place: Massively Multiplayer Online Gaming in American Youth Culture," *Tidskrift Journal of Research in Teacher Education* 3 (2005): 17–32; T. L. Taylor, "Beyond Management: Considering Participatory Design and Governance in Player Culture," *First Monday* 11 (2006).

23. John Dewey, *The Child and the Curriculum and the School and Society* (Chicago: University of Chicago Press, 1900 [1956]).

24. For research on heavy Internet use and face-to-face contact, see Norman Nie and Lutz Erbring, "Internet and Society." For research on distinctions between those who play games at home and work, see Norman Nie and D. Sunshine Hillygus, "The Impact of Internet Use on Sociability: Time-Diary Findings," *IT and Society* 1 (2002): 1–20.

25. Robert Putnam, *Bowling Alone* (New York: Simon and Schuster, 2000), 104.

26. Dmitri Williams, "The Impact of Time Online: Social Capital and Cyberbalkanization," *CyberPsychology and Behavior* 8 (2007): 580–584.

27. Chad Raphael et al., "Games for Civic Learning: A Conceptual Framework and Agenda for Research and Design" (unpublished manuscript, Santa Clara University, Santa Clara, CA); Kurt Squire and Henry Jenkins, "Harnessing the Power of Games in Education," *Insight* 3, no. 5 (2003): 7–33; Mary Flanagan, "Making Games for Social Change," *AI and Society: The Journal of Human-Centered Systems* 20, no. 1 (2006): 493–505; Celia Pearce et al., "Sustainable Play: Towards a New Games Movement for the Digital Age," *Games and Culture*, no. 3 (2007): 261–278.

28. Daniel Hart et al., "High School Community Service as a Predictor of Adult Voting and Volunteering," *American Educational Research Journal* 44 (2007): 197–219; Elizabeth S. Smith, "The Effects of Investment in the Social Capital of Youth on Political and Civic Behavior in Young Adulthood: A Longitudinal Analysis," *Political Psychology* 20 (1999): 553–580; Daniel A. McFarland and Reuben J. Thomas, "Bowling Young: How Youth Voluntary Associations Influence Adult Political Participation," *American Sociological Review* 71 (2006): 401–425; James Youniss and Miranda Yates, "Community Service and Political Identity"; Joseph Kahne, Bernadette Chi, and Ellen Middaugh, "Building Social Capital for Civic and Political Engagement: The Potential of High School Government Courses," *Canadian Journal of Education* 29 (2006): 387–409; Judith Torney-Purta, "The School's Role in Developing Civic Engage-

ment: A Study of Adolescents in Twenty-Eight Countries," *Applied Developmental Science* 6, no. 4 (2002): 203–212; Richard Niemi and Jane Junn, *Civic Education* (New Haven, CT: Yale University Press, 1998); Michael McDevitt and Spiro Kiousis, "Education for Deliberative Democracy: The Long-term Influence of Kids Voting USA" (CIRCLE working paper no. 22, Center for Information and Research on Civic Learning and Engagement, New York, 2004); Edward Metz and James Youniss, "Longitudinal Gains in Civic Development through School-based Required Service," *Political Psychology* 26 (2005): 413–438; Joseph Kahne and Susan Sporte, "Developing Citizens: The Impact of Civic Learning Opportunities on Students' Commitment to Civic Participation," *American Educational Research Journal* 45 (2008): 738–766; for review, see Gibson and Levine, *The Civic Mission of Schools.*

29. Gibson and Levine, *The Civic Mission of Schools*; Kahne, Chi, and Middaugh, "Building Social Capital."

30. James Youniss and Miranda Yates, *Community Service and Social Responsibility in Youth* (Chicago: University of Chicago Press, 1997).

31. Ibid.

32. Gibson and Levine, *The Civic Mission of Schools.*

33. See Simtropolis (Web site). Forum: *SimCity* 4, General Discussion, City Population Help, post 377, May 19, 2008. http://www.simtropolis .com/forum/messageview.cfm?catid=22&threadid=99623&enterthread=y.

34. Steven E. Finkel and Howard R. Ernst, "Civic Education in Post-Apartheid South Africa: Alternative Paths to the Development of Political Knowledge and Democratic Values," *Political Psychology* 26 (2005): 339. Reviews cited in Finkel and Ernst that make this point include: Martin Fishbein and Icek Ajzen, *Belief, Attitude, Intention and Behavior: An Introduction to Theory and Research* (Reading, MA: Addison-Wesley, 1975); and Philip G. Zimbardo and Michael R. Leippe, *The Psychology of Attitude Change and Social Influence* (Philadelphia: Temple University Press, 1991).

35. Kurt Squire and Sasha Barab, "Replaying History: Engaging Urban Underserved Students in Learning World History through Computer Simulation Games," in *Proceedings of the Sixth International Conference of the Learning Sciences,* ed. Y. B. Kafai et al. (Mahwah, NJ: Lawrence Erlbaum, 2004), 505–512.

36. Sasha Barab et al., "The Quest Atlantis Project: A Socially Responsive Play Space for Learning," in *The Educational Design and Use of Simulation Computer Games,* ed. B. E. Shelton and D. Wiley (Rotterdam, The Netherlands: Sense Publishers, 2007).

37. Ibid.

38. Elizabeth S. Smith, "The Effects of Investment in the Social Capital of Youth on Political and Civic Behavior in Young Adulthood: A Longitudinal Analysis," *Political Psychology* 20 (1999): 553–580.

39. Jenkins, "Confronting the Challenges."

40. Thomas and Brown, "Why Virtual Worlds Can Matter," 4.

41. Constance Steinkuehler and Dmitri Williams, "Where Everybody Knows Your (Screen) Name: Online Games as 'Third Places,'" *Journal of Computer-Mediated Communication* 11 (2006): 885–909.

42. Henry Jenkins, "MIT's Jenkins, Author Johnson, Talk Community, Creativity." Interview, Worlds in Motion.biz. http://www.worldsinmotion.biz/2008/03/mits_jenkins_author_johnson_ta.php (retrieved September 2, 2008).

43. Dmitri Williams, "Groups and Goblins: The Social and Civic Impact of Online Gaming," *Journal of Broadcasting and Electronic Media* 50, no. 4 (2006): 651–670.

44. Ito et al., *Hanging Out,* 220.

45. Williams, "The Impact of Time Online."

46. Pippa Norris, *Digital Divide: Civic Engagement, Information Poverty, and the Internet Worldwide* (Cambridge: Cambridge University Press,

2001); Bruce Bimber, *Information and American Democracy: Technology in the Evolution of Political Power* (Cambridge: Cambridge University Press, 2003); Karen Mossberger, Caroline J. Tolbert, and Ramona S. McNeal, *Digital Citizenship: The Internet, Society, and Participation* (Cambridge, MA: MIT Press, 2008).

47. Mossberger, Tolbert, and McNeal, *Digital Citizenship*.

48. Joseph Kahne and Ellen Middaugh, "Democracy for Some: The Civic Opportunity Gap in High School" (CIRCLE working paper no. 59, Center for Information and Research on Civic Learning and Engagement, New York, 2008).

49. The questions used to assess "civic gaming experiences" were based on classroom-based experiences civic education research has found to promote civic and political engagement in young people (see Gibson and Levine, *The Civic Mission of Schools*) and the gaming experiences that digital media scholars have proposed may support civic engagement (for example, Jenkins, "Confronting the Challenges").

50. In our analysis of civic gaming experiences, we excluded young people who answered, "does not apply." This response could be interpreted as meaning "never" having had this experience while playing games, but it could also be interpreted in other ways, so we did not include those responses. To ensure that excluding those young people did not alter our findings, we also ran our analysis using the alternate coding system, recoding "does not apply" answers as "never," and found a very similar set of relationships. The main difference is that two of our relationships that approached statistical significance, and therefore were not noted, became statistically significant. Specifically, the teens with the most civic game experiences were more likely to volunteer and teens with some civic gaming experiences were more likely to protest compared to teens with the least civic gaming experiences. Our overall conclusions are not affected.

51. Although these relationships are consistent and statistically significant, the overall impact of civic gaming experiences on civic outcomes

does not explain a high percentage of the overall variation in civic and political engagement (this is indicated by the R^2 in the tables in appendix B). This is not surprising as we do not expect that video game play is a prime determinant of civic and political engagement.

52. Steinkuehler and Williams, "Where Everybody Knows Your (Screen) Name."

53. We considered doing a similar analysis assessing the associations between playing games in the 14 gaming genres and civic gaming experiences. Several factors limit our confidence in such an analysis. For example, the analysis would introduce a large number of new independent variables, these independent variables are often highly correlated, the genres were not designed to group games in relation to the civic learning opportunities they provide, and it would be difficult to know which games within the genres might be responsible for any association that was identified.

54. Lenhart et al., "Teens and Video Games."

55. Seth Schiesel, "Exploring Fantasy Life and Finding a $4 Billion Franchise," *New York Times*, April 16, 2008. http://www.nytimes.com/2008 /04/16/arts/television/16sims.html?_r=1&scp=2&sq=Electronic%20 Arts%20Sims%20100%20million&st=cse&oref=slogin (retrieved July 10, 2008).

56. Lenhart et al., "Teens and Video Games."

57. Kahne, Chi, and Middaugh, "Building Social Capital"; Kahne and Sporte, "Developing Citizens"; McDevitt and Kiousis, "Education for Deliberative Democracy"; Metz and Youniss, "Longitudinal Gains in Civic Development"; Kenneth P. Langton and M. Kent Jennings, "Political Socialization and the High School Civic Curriculum in the United States," *American Political Science Review* 62 (1968): 862–867.

58. Gibson and Levine, *The Civic Mission of Schools*; Shelley Billig, "Research on K–12 School-Based Service-Learning: The Evidence Builds," *Phi Delta Kappan* 81 (2000): 658–664.

59. We did find a marginally significant relationship ($p \leq .10$) between guild membership and two of the eight indicators of civic engagement.

60. Putnam, *Bowling Alone*.

61. McFarland and Thomas, "Bowling Young."

62. Sidney Verba, Kay L. Schlozman, and Henry E. Brady, *Voice and Equality: Civic Voluntarism in American Politics* (Cambridge, MA: Harvard University Press, 1995).

63. Peter Levine, *The Future of Democracy: Developing the Next Generation of American Citizens* (Medford, MA: Tufts University Press, 2007).

64. Mossberger, Tolbert, and McNeal, *Digital Citizenship*.

65. Kahne and Middaugh, "Democracy for Some."

66. Erica W. Austin, "Exploring the Effects of Active Parental Mediation in Television Content," *Journal of Broadcasting and Electronic Media* 37 (1993): 147.

67. Molly W. Andolina et al., "Habits from Home, Lessons from School: Influences on Youth Civic Development," *PS: Political Science and Politics* 36, no. 2 (2003): 275–280; Hugh McIntosh, Daniel Hart, and James Youniss, "The Influence of Family Political Discussion on Youth Civic Development: Which Parent Qualities Matter?" *PS: Political Science and Politics* (July 2007).

68. Umashi Bers Marina and Clement Chau, "Fostering Civic Engagement by Building a Virtual City," *Journal of Computer-Mediated Communication* 11 (2006): 748–770.

69. Sarah E. Long, *The New Student Politics: The Wingspread Statement on Student Civic Engagement* (Providence, RI: Campus Compact, 2002).

70. Molly W. Andolina et al., "Searching for the Meaning of Youth Civic Engagement: Notes from The Field," *Applied Developmental Science* 6, no. 4 (2002): 189–195.

71. Gibson and Levine, *The Civic Mission of Schools*.

72. Jenkins, "Confronting the Challenges."

73. John W. Rice, "New Media Resistance: Barriers to Implementation of Computer Video Games in the Classroom," *Journal of Educational Multimedia and Hypermedia* 16, no. 3 (2007): 249–261.

74. Kahne and Sporte, "Developing Citizens."

75. Mary Flanagan and Helen Nissenbaum, "A Game Design Methodology to Incorporate Social Activist Themes," *Proceedings of CHI 2007* (New York: ACM Press, 2007), 181–190.

76. Ito et al., *Hanging Out.*

77. Erikson, *Identity;* Youniss and Yates, *Community Service and Social Responsibility;* Joseph Kahne and Joel Westheimer, "Teaching Democracy: What Schools Need To Do," *Phi Delta Kappan* 85, no. 1 (2003): 34–40, 57–66; Joseph Kahne and Ellen Middaugh, "High Quality Civic Education: What It Is and Who Gets It," *Social Education* 72, no. 1 (2008): 34–39.

78. Stephane Baldi et al., *What Democracy Means to Ninth Graders: U.S. Results from the International IEA Civic Education Study* (Washington, DC: U.S. Department of Education, National Center for Education Statistics, 2001).

79. John W. Rice, "New Media Resistance"; James Paul Gee, "Games and Learning: Issues, Perils, and Potentials: A Report to the Spencer Foundation," Spencer Foundation Report (Chicago: Spencer Foundation, 2006); Richard Halverson, "What Can K–12 School Leaders Learn from Video Games and Gaming?" *Innovate* 1, no. 6 (2005).

80. Verba, Schlozman, and Brady, *Voice and Equality;* Paul R. Abramson and William Claggett, "Recruitment and Political Participation," *Political Research Quarterly* 54 (2001): 905–916.

81. We are currently looking at this relationship in a related panel study.

82. Bill Bigelow Kolko and Marta Larsen, "On the Road to Cultural Bias: 'The Oregon Trail,'" *Equity Coalition* 5 (1999): 22–25; Kevin Schut, "Strategic Simulations and our Past: The Bias of Computer Games in the Presentation of History," *Games and Culture* 2 (2007): 213–235; Squire and Jenkins, "Harnessing the Power of Games."

83. Nick Yee, "Playing Politics: Videogames for Politics, Activism, and Advocacy," *First Monday* 11, no. 9 (2006): *Command Lines: The Emergence of Governance in Global Cyberspace;* Douglas Thomas, "KPK, Inc.: Race, Nation, and Emergent Culture in Online Games," in *Learning Race and Ethnicity: Youth and Digital Media,* ed. Anna Everett (Cambridge, MA: MIT Press, 2008), 155–174.

84. Raphael et al., "Games for Civic Learning."

85. Curt Feldman, "*Grand Theft Auto IV* Steals Sales Records," CNN online, May 8, 2008). http://www.cnn.com/2008/TECH/05/08/gta.sales/index.html (retrieved September 2008).

86. "*Dark Knight* Sets Weekend Box Office Record," CNN.com, July 20, 2008. http://www.cnn.com/2008/SHOWBIZ/Movies/07/20/dark.knight.ap/;http://www.suntimes.com/entertainment/movies/1077194, CST-FTR-box28.article.

87. In households with more than one 12- to 17-year-old, interviewers asked parents about, and conducted interviews with, a child selected at random.

88. PSRAI's disposition codes and reporting are consistent with the American Association for Public Opinion Research standards.

89. PSRAI assumes that 75 percent of cases that result in a constant disposition of "No answer" or "Busy" are actually not working numbers.

90. Parent education was also measured as a proxy for SES. We ran parallel analyses substituting this measure for income, and found some small differences in model fit and parameter estimates, but not substantial enough differences to choose one measure over the other.